Undergraduate Topics in Computer Science

'Undergraduate Topics in Computer Science' (UTiCS) delivers high-quality instructional content for undergraduates studying in all areas of computing and information science. From core foundational and theoretical material to final-year topics and applications, UTiCS books take a fresh, concise, and modern approach and are ideal for self-study or for a one- or two-semester course. The texts are all authored by established experts in their fields, reviewed by an international advisory board, and contain numerous examples and problems, many of which include fully worked solutions.

More information about this series at http://www.springer.com/series/7592

Gerard O'Regan

Introduction to the History of Computing

A Computing History Primer

Springer

Gerard O'Regan
Mallow, CK, Ireland

ISSN 1863-7310 ISSN 2197-1781 (electronic)
Undergraduate Topics in Computer Science
ISBN 978-3-319-33137-9 ISBN 978-3-319-33138-6 (eBook)
DOI 10.1007/978-3-319-33138-6

Library of Congress Control Number: 2016938793

Printed on acid-free paper

This Springer imprint is published by Springer Nature
The registered company is Springer International Publishing AG Switzerland

To
Pilar Cuitino Tride
In memory
of a beautiful week in Valparaiso

Preface

Overview

The objective of this book is to provide a concise introduction to the history of computing. The computing field is a vast area and a comprehensive account of its history would require several volumes. The goals of this book are more modest, and it aims to give the reader a flavour of some of the important events in the history of computing and to stimulate the reader to study the more advanced articles and books that are available.

Organization and Features

The first chapter provides an introduction to analog and digital computers and the von Neumann architecture which is the fundamental architecture underlying a digital computer. Chapter 2 considers the contributions of early civilizations to the computing field, and we discuss the achievements of the Babylonians, Egyptians, Greeks and Romans and the Islamic civilization.

Chapter 3 provides an introduction to the foundations of computing, and we discuss the binary number system and the step reckoner calculating machine, which were invented by Leibniz. Babbage designed the difference engine as a machine to evaluate polynomials, and his analytic engine provided the vision of a modern computer. Boole was an English mathematician who made important contributions to mathematics and logic, and Boole's symbolic logic provided the foundation for digital computing.

Chapter 4 discusses the first digital computers including the Atanasoff-Berry computer developed in the United States, the ENIAC and EDVAC computers developed in the United States, the Colossus computer developed in England, Zuse's computers developed in Germany and the Manchester Mark I computer developed in England.

Chapter 5 discusses the first commercial computers including UNIVAC developed by EMCC/Sperry in the United States, the LEO I computer developed by J. Lyons and Co. in England, the Z4 computer developed by Zuse KG in Germany and the Ferranti Mark I computer developed by Ferranti in England.

Chapter 6 discusses early commercial computers including the IBM 701 and 704 computers. We discuss the SAGE air defence system, which used the AN/FSQ-7 computer, which was developed by IBM. We discuss the invention of the transistor by William Shockley and others at Bell Labs and early transistor computers.

Chapter 7 discusses the invention of the integrated circuit by Jack Kilby at Texas Instruments and subsequent work by Robert Noyce at Fairchild Semiconductors on silicon-based integrated circuits. Moore's law on the exponential growth of transistor density on an integrated circuit is discussed, as well as its relevance to the computing power of electronic devices.

Chapter 8 is concerned with the development of the IBM System/360 and its influence on later computer development. The System/360 was a family of mainframe computers, and the user could start with a low specification member of the family and upgrade over time to a more powerful member of the family. It was the start of an era of computer compatibility, and it set IBM on the road to dominate the computing field for the next 20 years. It was a massive $5 billion gamble by IBM, and it moved the company from its existing product lines to the unknown world of the System/360.

Chapter 9 discusses later mainframes and minicomputers, including DEC's PDP-1, PDP-11 and VAX-11/780 minicomputers, which were popular with the engineering and scientific communities. We discuss Amdahl's mainframe computers such as the Amdahl 470 V/6 and the intense competition between IBM and Amdahl. DEC became the second largest computer company in the world in the late 1980s, but it was too slow in reacting to the rise of the microprocessor and the revolution in home computers.

Chapter 10 is concerned with the revolutionary invention of the microprocessor and discusses early microprocessors such as the Intel 4004, the 8-bit Intel 8080 and the 8-bit Motorola 6800. The 16-bit Intel 8086 was introduced in 1978, and the 16/32-bit Motorola 68000 was released in 1979. The 8-bit Intel 8088 (the cheaper 8-bit variant of the Intel 8086) was introduced in 1979, and it was chosen as the microprocessor for the IBM personal computer.

Chapter 11 discusses home computers such as the Apple I and II home computers, which were released in 1976 and 1977, respectively. We discuss the Commodore PET computer, which was introduced in 1977, and the Atari 400 and 800 computers, which were released in 1979. The Commodore 64 computer became very popular after its introduction in 1982. The Sinclair ZX 81 and ZX spectrum computers were released in 1980 and 1981, respectively, and the Apple Macintosh was released in 1984.

Chapter 12 discusses the introduction of the IBM personal computer, which was a major milestone in the computing field. IBM's goal was to get into the home computer market as quickly as possible, and this led IBM to build the machine from off-the-shelf parts from a number of equipment manufacturers. IBM outsourced the development of the operating system to a small company called Microsoft, and Intel was chosen to supply the microprocessor for the IBM PC. The open architecture of the IBM PC led to a new industry of IBM compatible computers.

Chapter 13 presents a short history of telecommunications, and it focuses on the development of mobile phone technology. The development of the AXE system by Ericsson is discussed, and this was the first fully automated digital switching system. We discuss the concept of a cellular system, which was introduced by Bell Labs, as well as the introduction of the first mobile phone, the DynaTAC, by Motorola.

Chapter 14 describes the Internet revolution starting from ARPANET, which was a packet-switched network, to TCP/IP, which is a set of network standards for interconnecting networks and computers. These developments led to the birth of the Internet, and Tim Berners-Lee's work at CERN led to the birth of the World Wide Web. Applications of the World Wide Web are discussed, as well as successful and unsuccessful new technology companies. The dot-com bubble and subsequent burst of the late 1990s/early 2000s are discussed.

Chapter 15 discusses the invention of the smartphone and the rise of social media. It describes the evolution of the smartphone from PDAs and mobile phone technology, and a smartphone is essentially a touch-based computer on a phone. The impact of Facebook and Twitter in social networking is discussed. Facebook is the leading social media site in the world, and it has become a way for young people to discuss their hopes and aspirations as well as a tool for social protest and revolution. Twitter has become a popular tool in political communication, and it is also an effective way for businesses to advertise its brand to its target audience.

Chapter 16 presents a short history of programming languages, starting from machine languages to assembly languages, to early high-level procedural languages such as FORTRAN and COBOL, to later high-level languages such as Pascal and C and to object-oriented languages such as C++ and Java. Functional programming languages and logic programming languages are discussed, and there is a short discussion on the important area of syntax and semantics.

Chapter 17 presents a short history of operating systems including the IBM OS/360, which was the operating system for the IBM System/360 family of computers. We discuss the MVS and VM operating systems, which were used on the IBM System/370 mainframe computer. Ken Thompson and Dennis Ritchie developed the popular UNIX operating system in the early 1970s. This is a multi-user and multitasking operating system and was written almost entirely in C. DEC developed the VAX/VMS operating system in the late 1970s for its VAX family of minicomputers. Microsoft developed MS/DOS for the IBM personal computer in 1981, and it introduced Windows as a response to the Apple Macintosh. There is a short discussion on Android and *i*OS, which are popular operating systems for mobile devices.

Chapter 18 presents a short history of software engineering from its birth at the Garmisch conference in Germany. The IEEE definition of software engineering is discussed, and it is emphasized that software engineering is a lot more than just programming. We discuss the key challenges in software engineering, as well as a number of the high-profile software failures. The waterfall and spiral lifecycles are discussed, as well a brief discussion on the Rational Unified Process and the popular

Agile methodology. We discuss the key activities in the waterfall model such as requirements, design, implementation, unit, system and acceptance testing.

Chapter 19 presents a short history of artificial intelligence, and we discuss the Turing test, which is a test of machine intelligence, describing it as strong and weak AI, where strong AI considers an AI programmed computer to be essentially a mind, whereas weak AI considers a programmed computer as simulating thought without real understanding. We discuss Searle's Chinese room argument, which is a rebuttal of strong AI. We discuss Weizenbaum's views on the ethics of AI and philosophical issues in AI. We discuss logic, neural networks and expert systems.

Chapter 20 presents a short history of databases including a discussion of the hierarchical and network models. We discuss the relational model as developed by Codd at IBM in more detail, as most databases used today are relational. There is a short discussion on the SQL query language and on the Oracle database.

Audience

The main audience of this book is computer science students who are interested in learning about the history of computing field. The book will also be of interest to the general reader who is curious about the history of computing.

Acknowledgements

I am deeply indebted to friends and family who supported my efforts in this endeavour. I would like to pay a special thanks to Pilar Cuitino Tride (to whom this book is dedicated). It was my privilege to meet her and to share special times with her in Valparaiso.

I would like to express my thanks to the team at Springer for their consistent professional work and a special thanks to Wayne Wheeler and Simon Rees.

I would like to thank all copyright owners for the permission to use their images. I believe that all of the required permissions have been obtained, but in the unlikely event that an image has been used without the appropriate authorization, please contact me so that the required permission can be obtained.

Cork, Ireland Gerard O'Regan

Contents

List of Figures

List of Tables

What Is a Computer?

1

Abstract

This chapter provides an introduction to computing, and a computer is a programmable electronic device that can process, store and retrieve data. It processes data according to a set of instructions (or program), and all computers consist of two basic parts, namely, *hardware* and *software*. There are two distinct families of computing devices, namely, *digital computers* and the historical *analog computer*. These two types of computer operate on quite different principles, and the earliest computers were analog. We discuss the von Neumann architecture, which is the fundamental architecture underlying a digital computer.

Key Topics

Analog computers
Digital computers
Vacuum tubes
Transistors
Integrated circuits
Von Neumann architecture
Generation of computers
Hardware
Software

1.1 Introduction

Computers are an integral part of modern society and new technology has transformed the modern world into a global village. Communication today is conducted using text messaging, mobile phones, video calls over the Internet, email and social

G. O'Regan, *Introduction to the History of Computing*, Undergraduate Topics in Computer Science, DOI 10.1007/978-3-319-33138-6_1

media sites such as Facebook. New technology allows people to keep in touch with friends and family around the world, and the World Wide Web allows businesses to compete in a global market.

A computer is a programmable electronic device that can process, store and retrieve data. It processes data according to a set of instructions or program. All computers consist of two basic parts, namely, *hardware* and *software*. The hardware is the physical part of the machine, and the components of a digital computer include memory for short-term storage of data or instructions, an arithmetic/logic unit for carrying out arithmetic and logical operations, a control unit responsible for the execution of computer instructions in memory and peripherals that handle the input and output operations. Software is a set of instructions that tells the computer what to do.

The original meaning of the word *computer* referred to someone who carried out calculations rather than an actual machine. The early digital computers built in the 1940s and 1950s were enormous machines consisting of thousands of vacuum tubes. They typically filled a large room, but their computational power was a fraction of the personal computers used today.

There are two distinct families of computing devices, namely, *digital computers* and the historical *analog computer*. The earliest computers were analog not digital, and these two types of computer operate on quite different principles.

The computation in a digital computer is based on binary digits, i.e. '0' and '1'. Electronic circuits are used to represent binary numbers, with the state of an electrical switch (i.e. 'on' or 'off') representing a binary digit internally within a computer.

A digital computer is a sequential device that generally operates on data one step at a time. The data are represented in binary format, and a single transistor is used to represent a binary digit in a digital computer. Several transistors are required to store larger numbers. The earliest digital computers were developed in the 1940s.

An *analog computer* operates in a completely different way to a digital computer. The representation of data in an analog computer reflects the properties of the data that are being modelled. For example, data and numbers may be represented by physical quantities such as electric voltage in an analog computer, whereas a stream of binary digits represents them in a digital computer.

1.2 Analog Computers

James Thompson (who was the brother of the physicist Lord Kelvin) did early foundational work on analog computation in the nineteenth century. He invented a wheel and disc integrator, which was used in mechanical analog devices, and he worked with Kelvin to construct a device to perform the integration of a product of two functions. Kelvin later described a general-purpose analog machine (he did not build it) for integrating linear differential equations of any order. He built a tide-predicting analog computer that remained in use at the Port of Liverpool up to the 1960s.

Fig. 1.1 Vannevar Bush with the differential analyser

The operations in an analog computer are performed in parallel, and they are useful in simulating dynamic systems. They have been applied to flight simulation, nuclear power plants and industrial chemical processes.

Vannevar Bush at the Massachusetts Institute of Technology developed the first large-scale general-purpose mechanical analog computer. This machine was Bush's differential analyser (Fig. 1.1), and it was a mechanical analog computer designed to solve 6th-order differential equations by integration, using wheel-and-disc mechanisms to perform the integration. This mechanization allowed integration and differential equation problems to be solved more rapidly. The machine took up the space of a large table in a room and weighed 100 t.

It contained wheels, discs, shafts and gears to perform the calculations. It required a considerable set-up time by technicians to solve a particular equation. It contained 150 motors and miles of wires connecting relays and vacuum tubes.

Data representation in an analog computer is compact, but it may be subject to corruption with noise. A single capacitor can represent one continuous variable in an analog computer, whereas several transistors are required in a digital computer. Analog computers were replaced by digital computers shortly after the Second World War.

1.3 Digital Computers

Early digital computers used vacuum tubes to store binary information, and a vacuum tube could represent the binary value '0' or '1'. These tubes were large and bulky and generated a significant amount of heat. Air conditioning was required to cool the machine, and there were problems with the reliability of the tubes.

Shockley and others invented the transistor in the later 1940s, and it replaced vacuum tubes from the late 1950s. Transistors are small and consume very little power, and the resulting machines were smaller, faster and more reliable.

Integrated circuits were introduced in the early 1960s, and a massive amount of computational power could now be placed on a very small chip. Integrated circuits are small and consume very little power and may be mass-produced to very high-quality standard. However, integrated circuits are difficult to modify or repair and nearly always need to be replaced.

The fundamental architecture of a computer has remained basically the same since von Neumann and others proposed it in the mid-1940s. It includes a central processing unit which includes the control unit and the arithmetic unit, an input and output unit and memory.

1.3.1 Vacuum Tubes

A vacuum tube is a device that relies on the flow of an electric current through a vacuum. Vacuum tubes (thermionic valves) were widely used in electronic devices such as televisions, radios and computers until the invention of the transistor.

The basic idea of a vacuum tube is that a current passes through the filament, which then heats it up so that it gives off electrons. The electrons are negatively charged and are attracted to the small positive plate (or anode) within the tube. A unidirectional flow is thus established between the filament and the plate. Thomas Edison had observed this while investigating the reason for breakage of lamp filaments. He noted an uneven blackening (darkest near one terminal of the filament) of the bulbs in his incandescent lamps and noted that current flows from the lamp's filament and a plate within the vacuum.

The first generation of computers used several thousand bulky vacuum tubes, with several racks of vacuum tubes taking up the space of a large room. The vacuum tube used in the early computers was a three-terminal device, and it consisted of a cathode, a grid and a plate. The vacuum tube was used to represent one of two binary states: i.e. the binary value '0' or '1'.

The filament of a vacuum tube becomes unstable over time. In addition, if air leaks into the tube, then oxygen will react with the hot filament and damage it. The size and unreliability of vacuum tubes motivated research into more compact and reliable technologies. This led to the invention of the transistor in the late 1940s.

The first generation of digital computers all used vacuum tubes: e.g. the Atanasoff-Berry computer (ABC) developed at the University of Iowa in 1942; the Colossus developed at Bletchley Park in 1944; ENIAC developed in the United States in the mid-1940s; UNIVAC I developed in 1951; Whirlwind developed in 1951; and the IBM 701 developed in 1953.

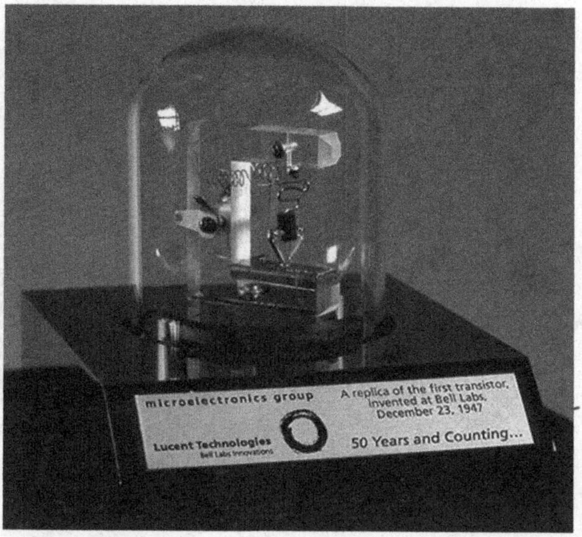

Fig. 1.2 Replica of transistor (Courtesy of Lucent Bell Labs)

1.3.2 Transistors

The transistor is a fundamental building block in modern electronic systems, and its invention revolutionized the field of electronics. It was smaller, cheaper and more reliable than the existing vacuum tubes.

The transistor is a three-terminal, solid-state electronic device. It can control electric current or voltage between two of the terminals by applying an electric current or voltage to the third terminal. The three-terminal transistor enables an electric switch to be made which can be controlled by another electrical switch. Complicated logic circuits may be built up by cascading these switches (switches that control switches that control switches and so on).

These logic circuits may be built very compactly on a silicon chip with a density of a million transistors per square centimetre. The switches may be turned on and off very rapidly (e.g. every 0.000000001 s). These electronic chips are at the heart of modern electron devices.

The transistor (Fig. 1.2) was developed at Bell Labs after the Second World War. The goal of the research was to find a solid-state alternative to vacuum tubes, as this technology was too bulky and unreliable. Three inventors at Bell Labs (Shockley, Bardeen and Brattain) were awarded the Nobel Prize in Physics in 1956 in recognition of their invention of the transistor.

William Shockley (Fig. 1.3) was involved in radar research and antisubmarine operations research during the Second World War, and after the war he led a research group including Bardeen and Brattain to find a solid-state alternative to the glass-based vacuum tubes.

Bardeen and Brattain succeeded in creating a point-contact transistor in 1947 independently of Shockley who was working on a junction-based transistor.

Fig. 1.3 William Shockley (Courtesy of Chuck Painter. Stanford News Service)

Shockley believed that the point-contact transistor would not be commercially viable, and his junction point transistor was announced in 1951.

Shockley was not an easy person to work with and relations between him and the others deteriorated. He formed Shockley Semiconductor Inc. (part of Beckman Instruments) in 1955.

The second generation of computers used transistors instead of vacuum tubes. The University of Manchester's experimental transistor computer was one of the earliest transistor computers. The prototype machine appeared in 1953 and the full-size version was commissioned in 1955. The invention of the transistor is discussed in more detail in Chap. 6.

1.3.3 Integrated Circuits

Jack Kilby of Texas Instruments invented the integrated circuit in 1958. His invention used a wafer of germanium, and Robert Noyce of Fairchild Semiconductors did subsequent work on silicon-based integrated circuits. The integrated circuit was a solution to the problem of building a circuit with a large number of components, and the Nobel Prize in Physics was awarded to Kirby in 2000 for his contribution to its invention.

The idea was that instead of making transistors one by one, several transistors could be made at the same time on the same piece of semiconductor. This allowed transistors and other electronic components such as resistors, capacitors and diodes to be made by the same process with the same materials.

An integrated circuit consists of a set of electronic circuits on a small chip of semiconductor material, and it is much smaller than a circuit made out of independent components. Integrated circuits today are extremely compact and may contain billions of transistors and other electronic components in a tiny area. The width of each conducting line has got smaller and smaller due to advances in technology over the years, and it is now measured in tens of nanometres.

The number of transistors per unit area has been doubling (roughly) every 1–2 years over the last 30 years. This amazing progress in circuit fabrication is known as Moore's law after Gordon Moore (one of the founders of Intel) who formulated the law in the mid-1960s [ORg:13].

Kilby was designing micromodules for the military, and this involved connecting many germanium[1] wafers of discrete components together by stacking each wafer on top of one another. The connections were made by running wires up the sides of the wafers.

Kilby saw this process as unnecessarily complicated and realized that if a piece of germanium was engineered properly, it could act as many components simultaneously. This was the idea that led to the birth of the first integrated circuit, and its development involved miniaturizing transistors and placing them on silicon chips called semiconductors. The use of semiconductors led to third-generation computers, with a major increase in speed and efficiency.

Users interacted with third-generation computers through keyboards and monitors and interfaced with operating systems, which allowed the device to run many different applications at one time with a central program that monitored the memory. Computers became accessible to a wider audience, as they were smaller and cheaper than their predecessors. The invention of the integrated circuit is discussed in more detail in Chap. 7.

1.3.4 Microprocessors

The Intel 4004 microprocessor (Fig. 1.4) was the world's first microprocessor, and it was released in 1969. It was the first semiconductor device that provided, at the chip level, the functions of a computer.

The invention of the microprocessor happened by accident rather than design. Busicom, a Japanese company, requested Intel to design a set of integrated circuits for its new family of high-performance programmable calculators. Ted Hoff, an Intel engineer, studied Busicom's design and rejected it as unwieldy. He proposed a more elegant solution requiring just four integrated circuits (Busicom's required 12 integrated circuits), and his design included a chip that was a general-purpose logic device that derived its application instructions from the semiconductor memory. This was the Intel 4004 microprocessor.

It provided the basic building blocks that are used in today's microcomputers, including the arithmetic and logic unit and the control unit. The 4-bit Intel 4004 ran at a clock speed of 108 kHz and contained 2300 transistors. It processed data in 4

[1]Germanium is an important semiconductor material used in transistors and other electronic devices.

Fig. 1.4 Intel 4004 microprocessor

bits, but its instructions were 8 bits long. It could address up to 1 Kb of program memory and up to 4 Kb of data memory.

Gary Kildall of Digital Research was one of the early people to recognize the potential of a microprocessor as a computer in its own right. He worked as a consultant with Intel, and he began writing experimental programs for the Intel 4004 microprocessor. He later developed the CP/M operating system for the Intel 8080 chip, and he set up Digital Research to market and sell the operating system.

The development of the microprocessor led to the fourth generation of computers with thousands of integrated circuits placed onto a single silicon chip. A single chip could now contain all of the components of a computer from the CPU and memory to input and output controls. It could fit in the palm of the hand, whereas first generation of computers filled an entire room. The invention of the microprocessor is discussed in more detail in Chap. 10.

1.4 Von Neumann Architecture

The earliest computers were fixed program machines that were designed to do a specific task. This proved to be a major limitation as it meant that a complex manual rewiring process was required to enable the machine to solve a different problem.

The computers used today are general-purpose machines designed to allow a variety of programs to be run on the machine. Von Neumann and others [VN:45] described the fundamental architecture underlying the computers used today in the late 1940s. It is known as von Neumann architecture (Fig. 1.5).

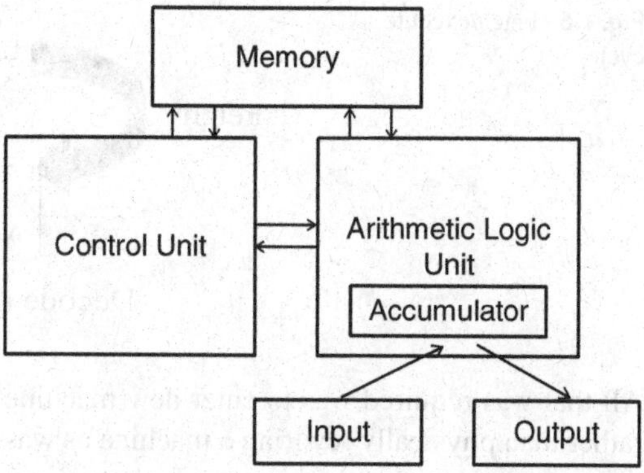

Fig. 1.5 Von Neumann architecture

Table 1.1 Von Neumann architecture

Component	Description
Arithmetic unit	The arithmetic unit is capable of performing basic arithmetic operations
Control unit	The program counter contains the address of the next instruction to be executed. This instruction is fetched from memory and executed. This is the basic fetch and execute cycle (Fig. 1.6)
	The control unit contains a built-in set of machine instructions
Input-output unit	The input and output unit allows the computer to interact with the outside world
Memory	The one-dimensional memory stores all of the program instructions and data. These are usually kept in different areas of memory
	The memory may be written to or read from: i.e. it is a random access memory (RAM)
	The program instructions are binary values, and the control unit decodes the binary value to determine the particular instruction to execute

Von Neumann architecture arose on work done by von Neumann, Eckert, Mauchly and others on the design of the EDVAC computer, which was the successor to ENIAC computer. Von Neumann's draft report on EDVAC [VN:45] described the new architecture.

Von Neumann architecture led to the birth of stored program computers, where a single store is used for both machine instructions and data. The key components of von Neumann architecture is described in Table 1.1.

The key approach to building a general-purpose device according to von Neumann was in its ability to store not only its data and the intermediate results of computation but also to store the instructions or commands for the computation. The computer instructions can be part of the hardware for specialized machines, but for general-purpose machines, the computer instructions must be as changeable as the data that is acted upon by the instructions. His insight was to recognize that both the machine instructions and data could be stored in the same memory.

The key advantage of the von Neumann architecture over the existing approach was that it was much simpler to reconfigure a computer to perform a different task.

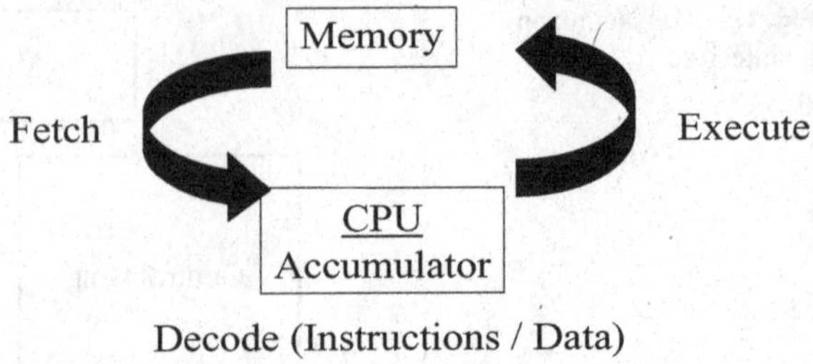

Fig. 1.6 Fetch/execute cycle

All that was required was to enter new machine instructions in computer memory rather than physically rewiring a machine as was required with ENIAC. The limitations of von Neumann architecture include that it is limited to sequential processing and not very suitable for parallel processing.

1.5 Hardware and Software

Hardware is the physical part of the machine. It is tangible and may be seen or touched. It includes punched cards, vacuum tubes, transistors and circuit boards, integrated circuits and microprocessors. The hardware of a personal computer includes a keyboard, network cards, a mouse, a DVD drive, hard disc drive, printers and scanners and so on.

Software is intangible and consists of a set of instructions that tells the computer what to do. It is an intellectual creation of a programmer or a team of programmers. Operating system software manages the computer hardware and resources and acts as an intermediary between the application programs and the computer hardware. Examples of operating systems include the OS/360 for the IBM System 360 mainframe, the UNIX operating system, the various Microsoft Windows operating system for the personal computer and the Mac operating system for the Macintosh computer.

1.6 Review Questions

1. Explain the difference between analog and digital computers.
2. Explain the difference between hardware and software.
3. What is a microprocessor?
4. Explain the difference between vacuum tubes, transistors and integrated circuits.
5. Explain von Neumann architecture.
6. What are the advantages and limitations of the von Neumann architecture?
7. Explain the difference between a fixed program machine and a stored program machine.

1.7 Summary

A computer is a programmable electronic device that can process, store and retrieve data. It processes data according to a set of instructions or program. All computers consist of two basic parts, namely, the *hardware* and *software*. The hardware is the physical part of the machine, whereas software is intangible and is the set of instructions that tells the computer what to do.

There are two distinct families of computing devices, namely, *digital computers* and the historical *analog computer*. These two types of computer operate on quite different principles. The earliest digital computers were built in the 1940s, and these were large machines consisting of thousands of vacuum tubes. However, their computational power was a fraction of what is available today.

A digital computer is a sequential device that generally operates on data one step at a time. The data are represented in binary format, and a single transistor in a digital computer can store only two states: i.e. on and off. Several transistors are required to store larger numbers.

The representation of data in an analog computer reflects the properties of the data that is being modelled. For example, data and numbers may be represented by physical quantities such as electric voltage in an analog computer. However, a stream of binary digits represents the data in a digital computer.

Computing in Early Civilizations

2

Abstract
This chapter considers the contributions of early civilizations to the computing field, including the achievements of the Babylonians, Egyptians, Greeks and Romans, and the Islamic world. The Babylonian civilization flourished in Mesopotamia (in modern Iraq) from about 2000 B.C. until about 300 B.C., and they made important contributions to mathematics. The Egyptian civilization developed along the Nile from about 4000 B.C., and their knowledge of mathematics allowed them to construct the pyramids at Giza as well as other impressive monuments. The Greeks made major contributions to Western civilization including mathematics, logic and philosophy. The golden age of Islamic civilization was from 750 A.D. to 1250 A.D., and during this period enlightened caliphs recognized the value of knowledge and sponsored scholars to come to Baghdad to gather and translate the existing world knowledge into Arabic.

Key Topics
Babylonian mathematics
Egyptian civilization
Greek and Roman civilization
Counting and numbers
Solving practical problems
Syllogistic logic, algorithms and early ciphers

G. O'Regan, *Introduction to the History of Computing*, Undergraduate Topics in Computer Science, DOI 10.1007/978-3-319-33138-6_2

2.1 Introduction

It is difficult to think of western society today without modern technology. The last decades of the twentieth century have witnessed a proliferation of high-tech computers, mobile phones, text messaging, the Internet and the World Wide Web. Software is now pervasive and it is an integral part of automobiles, airplanes, televisions and mobile communication. The pace of change is relentless, and communication today is instantaneous with text messaging, mobile phones and email. Today people may book flights over the World Wide Web as well as keeping in contact with family members in any part of the world. In previous generations, communication often involved writing letters that took months to reach the recipient. Communication improved with the telegraph and the telephone in the late nineteenth century, and the new generation probably views the world of their parents and grandparents as being old fashioned.

The new technologies have led to major benefits[1] to society and to improvements in the standard of living for many citizens in the western world. It has also reduced the necessity for humans to perform some of the more tedious or dangerous manual tasks, as computers may now automate many of these. The increase in productivity due to the more advanced computerized technologies has allowed humans, at least in theory, the freedom to engage in more creative and rewarding tasks.

Some early societies had a limited vocabulary for counting: e.g. 'one, two, three, many' is associated with a number of primitive societies and indicates limited numerate and scientific abilities. It suggests that the problems dealt with in this culture were elementary. These primitive societies generally employed their fingers for counting, and as humans have five fingers on each hand and five toes on each foot, then the obvious bases would have been 5, 10 and 20. Traces of the earlier use of the base 20 system are still apparent in modern languages such as English and French. This includes phrases such as *three score* in English and *quatre vingt* in French.

The decimal system (base 10) is used today in western society, but the base 60 was common in early computation *circa* 1500 B.C. One example of the use of base 60 today is still evident in the subdivision of hours into 60 min and the subdivision of minutes into 60 s. The base 60 system (i.e. the sexagesimal system) is inherited from the Babylonians [Res:84], and the Babylonians were able to represent arbitrarily large numbers or fractions with just two symbols.

The achievements of some of these early civilizations are impressive. The archaeological remains of ancient Egypt such as the pyramids at Giza and the temples of Karnak and Abu Simbel are inspiring. These monuments provide an indication of the engineering sophistication of the ancient Egyptian civilization. The

---

[1] Of course, while the new technologies are of major benefit to society, it is essential that the population of the world moves towards more sustainable development to ensure the long-term survival of the planet for future generations. This involves finding technological and other solutions to reduce greenhouse gas emissions as well as moving to a carbon neutral way of life. The solution to the environmental issues will be a major challenge for the twenty-first century.

objects found in the tomb of Tutankhamun[2] are now displayed in the Egyptian museum in Cairo and demonstrate the artistic skill of the Egyptians.

The Greeks made major contributions to western civilization including contributions to mathematics, philosophy, logic, drama, architecture, biology and democracy.[3] The Greek philosophers considered fundamental questions such as ethics, the nature of being, how to live a good life and the nature of justice and politics. The Greek philosophers include Parmenides, Heraclitus, Socrates, Plato and Aristotle. The Greeks invented democracy and their democracy was radically different from today's representative democracy.[4] The sophistication of Greek architecture and sculpture is evident from the Parthenon on the Acropolis and the Elgin marbles[5] that are housed today in the British Museum, London.

The Hellenistic[6] period commenced with Alexander the Great and led to the spread of Greek culture throughout most of the known world. The city of Alexandria became a centre of learning during the Hellenistic period. Its scholars included Euclid who provided a systematic foundation for geometry, and his famous work *The Elements* consists of 13 books.

There are many words of Greek origin that are part of the English language. These include words such as psychology which is derived from two Greek words *psyche* (ψυχη) and *logos* (λογος). The Greek word *psyche* means mind or soul, and the word *logos* means an account or discourse. Other examples are anthropology derived from *anthropos* (ανθροπος) and *logos* (λογος).

The Romans were influenced by Greek culture, and they built aqueducts, viaducts and amphitheatres. They also developed the Julian calendar, formulated laws

[2] Tutankhamun was a minor Egyptian pharaoh who reigned after the controversial rule of Akhenaten. Howard Carter discovered Tutankhamun's intact tomb in the Valley of the Kings. The quality of the workmanship of the artefacts found in the tomb was extraordinary, and a visit to the Egyptian museum in Cairo is memorable.

[3] The origin of the word 'democracy' is from demos (δημος) meaning people and kratos (κρατος) meaning rule. That is, it means rule by the people. It was introduced into Athens following the reforms introduced by Cleisthenes. He divided the Athenian city-state into 13 areas. Twenty of these areas were inland or along the coast and ten were in Attica itself. Fishermen lived mainly in the ten coastal areas; farmers in the ten inland areas; and various tradesmen in Attica. Cleisthenes introduced ten new clans where the members of each clan came from one coastal area and one inland area on one area in Attica. He then introduced a boule (or assembly) which consisted of 500 members (50 from each clan). Each clan ruled for $^1/_{10}$ of the year.

[4] The Athenian democracy involved the full participations of the citizens (i.e. the male adult members of the city-state who were not slaves), whereas in representative democracy the citizens elect representatives to rule and represent their interests. The Athenian democracy was chaotic and could also be easily influenced by individuals who were skilled in rhetoric. There were teachers (known as the sophists) who taught wealthy citizens rhetoric in return for a fee. The origin of the word 'sophist' is the Greek word σοφος meaning wisdom. One of the most well known of the sophists was Protagorus. The problems with Athenian democracy led philosophers such as Plato to consider alternate solutions such as rule by philosopher kings. This is described in Plato's Republic.

[5] The Elgin marbles are named after Lord Elgin who moved them from the Parthenon in Athens to London in 1806. The marbles show the Panathenaic festival that was held in Athens in honour of the goddess Athena after whom Athens is named.

[6] The origin of the word Hellenistic is from Hellene ('Ελλην) meaning Greek.

(*lex*) and maintained peace throughout the Roman Empire (*pax Romano*). The ruins of Pompeii and Herculaneum demonstrate their engineering excellence. Their numbering system is still employed in clocks and for page numbering in documents, but it is cumbersome for serious computation. The collapse of the Roman Empire in Western Europe led to a decline in knowledge and learning in Europe. However, the eastern part of the Roman Empire continued at Constantinople until the Ottomans conquered it in 1453.

2.2 The Babylonians

The Babylonian[7] civilization flourished in Mesopotamia (in modern Iraq) from about 2000 B.C. until about 300 B.C. Various clay cuneiform tablets containing mathematical texts were discovered and deciphered in the nineteenth century [Smi:23]. These included tables for multiplication, division, squares, cubes and square roots; measurement of area and length; and the solution of linear and quadratic equations. The late Babylonian period (c. 300 B.C.) includes work on astronomy.

The Babylonians recorded their mathematics on soft clay using a wedge-shaped instrument to form impressions of the *cuneiform* numbers. The clay tablets were then baked in an oven or by the heat of the sun. They employed just two symbols (1 and 10) to represent numbers, and these symbols were then combined to form all other numbers. They employed a positional number system[8] and used the base 60 system. The symbol representing 1 could also (depending on the context) represent 60, 60^2, 60^3, etc. It could also mean $^1/_{60}$, $^1/_{3600}$ and so on. There was no zero employed in the system and there was no decimal point (no 'sexagesimal point'), and therefore the context was essential.

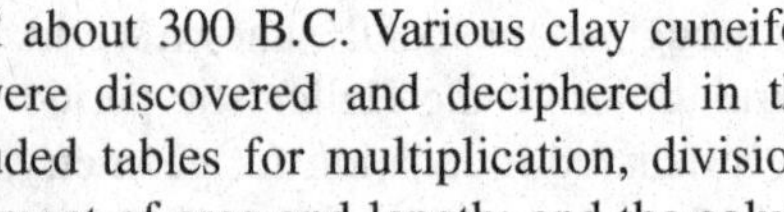

The example above illustrates the cuneiform notation and represents the number $60+10+1=71$. They used the base 60 system for computation, and one possible explanation for this is the ease of dividing 60 into parts as it is divisible by 2, 3, 4, 5, 6, 10, 12, 15, 20 and 30. They were able to represent large and small numbers and had no difficulty in working with fractions (in base 60) and in multiplying fractions. They maintained tables of reciprocals (i.e. $^1/n$, $n=1, \ldots, 59$ apart from numbers like 7, 11, etc., which are not of the form $2^\alpha 3^\beta 5^\gamma$ and cannot be written as a finite sexagesimal expansion).

[7] The hanging gardens of Babylon were one of the seven wonders of the ancient world.

[8] A positional numbering system is a number system where each position is related to the next by a constant multiplier. The decimal system is an example: e.g. $546=5*10^2+4*10^1+6$.

Babylonian numbers may be represented in a more modern sexagesimal notation [Res:84]. For example, 1;24,51,10 represents the number $1+{}^{24}/_{60}+{}^{51}/_{3600}+{}^{10}/_{216000}=1+0.4+0.0141666+0.0000462=1.4142129$ and is the Babylonian representation of the square root of 2. The Babylonians performed multiplication as the following calculation of (20) * (1;24,51,10), i.e. 20 * sqrt(2) illustrates:

$$20*1=20$$

$$20*;24=20*\frac{24}{60}=8$$

$$20*\frac{51}{3600}=\frac{51}{180}=\frac{17}{60}=;17$$

$$20*\frac{10}{216000}=\frac{3}{3600}+\frac{20}{216000}=;0,3,20$$

Hence, 20 * sqrt (2) = 20; + 8; +; 17 +; 0,3,20 = 28; 17, 3, 20.

The Babylonians appear to have been aware of Pythagoras' theorem about 1000 years before the time of Pythagoras. The Plimpton 322 tablet (Fig. 2.1) records various Pythagorean triples, i.e. triples of numbers (a, b, c) where $a^2+b^2=c^2$. It dates from approximately 1700 B.C.

They developed an algebra to assist with problem solving, which allowed problems involving length, breadth and area to be discussed and solved. They did not employ notation for representation of unknown values (e.g. let x be the length and y

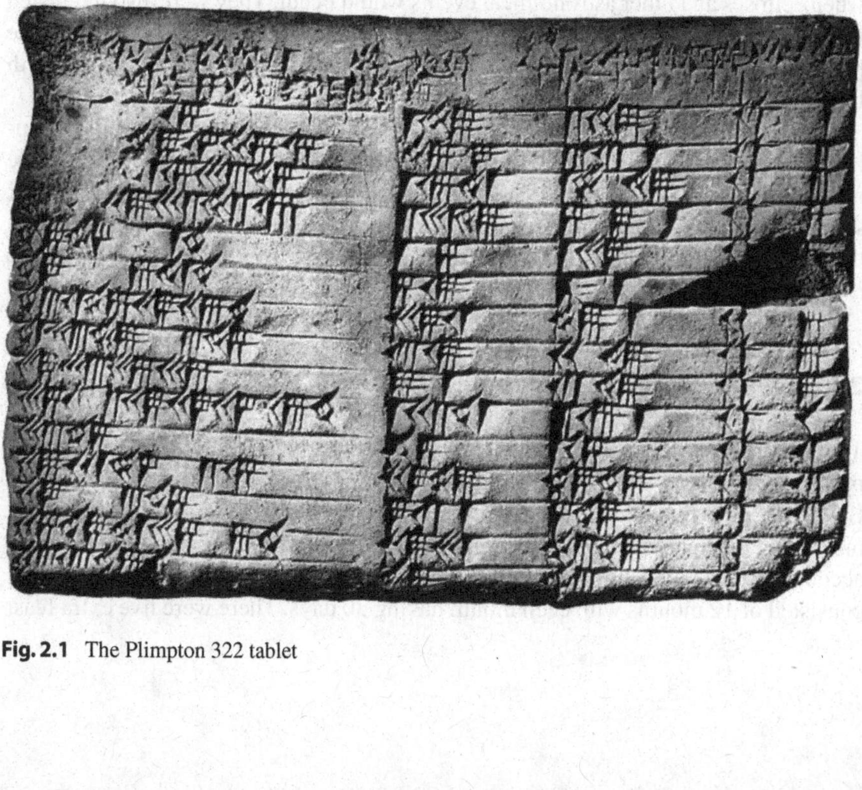

Fig. 2.1 The Plimpton 322 tablet

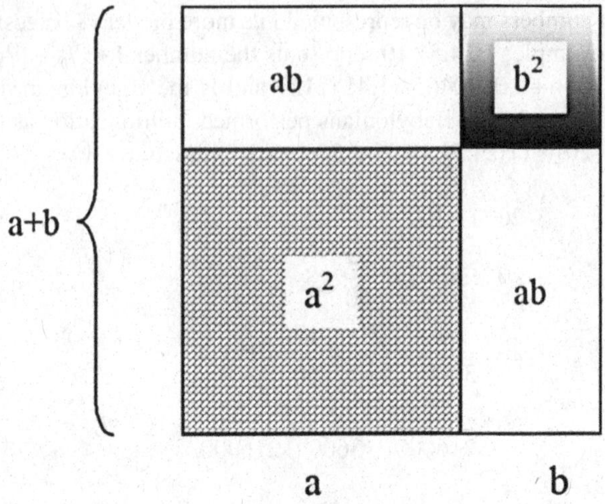

Fig. 2.2 Geometric representation of $(a+b)^2 = (a^2 + 2ab + b^2)$

be the breadth), and instead they used words like 'length' and 'breadth'. They were familiar with and used square roots in their calculations, as well as techniques to solve quadratic equations.

They were familiar with various mathematical identities such as $(a+b)^2 = (a^2 + 2ab + b^2)$ as illustrated geometrically in Fig. 2.2. They also worked on astronomical problems, and they had mathematical theories of the cosmos to predict when eclipses and other astronomical events would occur. They were also interested in astrology, and they associated various deities with the heavenly bodies such as the planets, as well as the sun and moon. Various clusters of stars were associated with familiar creatures such as lions, goats and so on.

The Babylonians used counting boards to assist with counting and simple calculations. A counting board is an early version of the abacus, and it was usually made of wood or stone. The counting board contained grooves, which allowed beads or stones to be moved along the groove. The abacus differed from counting boards in that the beads in abaci contained holes that enabled them to be placed in a particular rod of the abacus.

2.3 The Egyptians

The Egyptian civilization developed along the Nile from about 4000 B.C., and the pyramids at Giza were built during the fourth dynasty around 3000 B.C. The Egyptians used mathematics to solve practical problems such as measuring time, measuring the annual Nile flooding, calculating the area of land, book-keeping and accounting and calculating taxes. They developed a calendar circa 4000 B.C., which consisted of 12 months with each month having 30 days. There were five extra feast

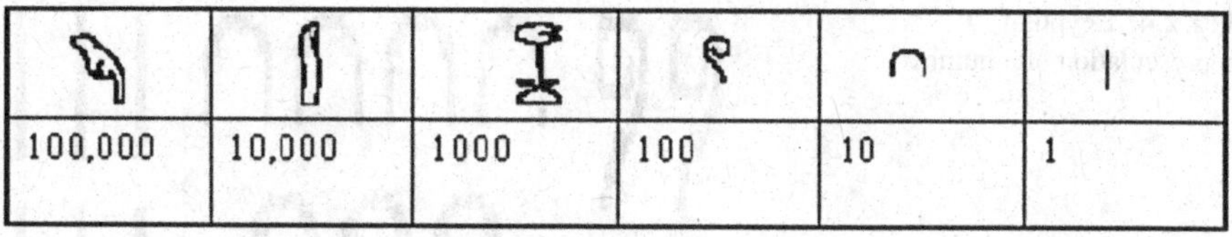

Fig. 2.3 Egyptian numerals

days to give 365 days in a year. Egyptian writing commenced around 3000 B.C., and it is recorded on the walls of temples and tombs.[9] A reedlike parchment termed 'papyrus' was also used for writing, and there are three Egyptian writing scripts, namely, the *hieroglyphics*, the *hieratic script* and the *demotic script*.

Hieroglyphs are little pictures and are used to represent words, alphabetic characters as well as syllables or sounds. Champollion deciphered hieroglyphics with his work on the Rosetta stone, which was discovered during the Napoleonic campaign in Egypt, and it is now in the British Museum in London. It contains three scripts: hieroglyphics, demotic script and Greek. A key part of the decipherment was that the Rosetta stone contained just one name 'Ptolemy' in the Greek text, and this was identified with the hieroglyphic characters in the cartouche[10] of the hieroglyphics. There was just one cartouche on the Rosetta stone, and Champollion inferred that the cartouche represented the name 'Ptolemy'. He was familiar with another multilingual object, which contained two names in the cartouche. One he recognized as Ptolemy and the other he deduced from the Greek text as 'Cleopatra'. This led to the breakthrough in translation of the hieroglyphics [Res:84], and Champollion's knowledge of Coptic was also essential in the deciphering

The Rhind papyrus is a famous Egyptian papyrus on mathematics. The Scottish Egyptologist, Henry Rhind, purchased it in 1858, and it is a copy created by an Egyptian scribe called Ahmose.[11] It is believed to date to 1832 B.C. It contains examples of all kinds of arithmetic and geometric problems, and students may have used it as a textbook to develop their mathematical knowledge. This would allow them to participate in the pharaoh's building programme.

The Egyptians were familiar with geometry, arithmetic and elementary algebra. They had formulae to find solutions to problems with one or two unknowns. A base 10 number system was employed with separate symbols for one, ten, a hundred, a thousand, a ten thousand, a hundred thousand and so on. These hieroglyphic symbols are represented in Fig. 2.3.

[9] The decorations of the tombs in the Valley of the Kings record the life of the pharaoh including his exploits and successes in battle.

[10] The cartouche surrounded a group of hieroglyphic symbols enclosed by an oval shape. Champollion's insight was that the group of hieroglyphic symbols represented the name of the Ptolemaic pharaoh 'Ptolemy'.

[11] The Rhind papyrus is sometimes referred to as the Ahmose papyrus in honour of the scribe who wrote it in 1832 B.C.

Fig. 2.4 Egyptian representation of a number

Fig. 2.5 Egyptian representation of a fraction

For example, the representation of the number 276 in Egyptian hieroglyphics is given in Fig. 2.4.

The addition of two numerals is straightforward and involves adding the individual symbols, and where there are ten copies of a symbol, it is then replaced by a single symbol of the next higher value. The Egyptian employed unit fractions (e.g. $1/n$ where n is an integer). These were represented in hieroglyphs by placing the symbol representing a 'mouth' above the number. The symbol 'mouth' represents part of. For example, the representation of the number $^{1}/_{276}$ is given in Fig. 2.5.

The papyrus included problems to determine the angle of the slope of the pyramid's face. The Egyptians were familiar with trigonometry including sine, cosine, tangent and cotangent, and they knew how to build right angles into their structures by using the ratio 3:4:5. The papyrus also considered problems such as the calculation of the number of bricks required for part of a building project. They were familiar with addition, subtraction, multiplication and division. However, their multiplication and division were cumbersome as they could only multiply and divide by two.

Suppose they wished to multiply a number n by 7. Then $n * 7$ is determined by $n * 2 + n * 2 + n * 2 + n$. Similarly, if they wished to divide 27 by 7, they would note that $7 * 2 + 7 = 21$ and that $27 - 21 = 6$ and that therefore the answer was $3\,^{6}/_{7}$. Egyptian mathematics was cumbersome and the writing of it was long and repetitive. For example, they wrote a number such as 22 by 10 + 10 + 1 + 1.

The Egyptians calculated the approximate area of a circle by calculating the area of a square $^{8}/_{9}$ of the diameter of a circle. That is, instead of calculating the area in terms of our familiar πr^2, their approximate calculation yielded $(^{8}/_{9} * 2r)^2 = {}^{256}/_{81}\, r^2$

or 3.16 r^2. Their approximation of π was $^{256}/_{81}$ or 3.16. They were able to calculate the area of a triangle and volumes.

The Moscow papyrus is a well-known Egyptian papyrus, and it includes a problem to calculate the volume of the frustum. The formula for the volume of a frustum of a square pyramid[12] was given by $V = {}^1/_3\, h(b_1^2 + b_1 b_2 + b_2^2)$, and when b_2 is 0, then the well-known formula for the volume of a pyramid is given, i.e. ${}^1/_3\, hb_1^2$.

2.4 The Greeks

The Greeks made major contributions to western civilization including mathematics, logic, astronomy, philosophy, politics, drama and architecture. The Greek world of 500 B.C. consisted of several independent city-states such as Athens and Sparta and various city-states in Asia Minor. The Greek polis (πολις) or city-state tended to be quite small, and it consisted of the Greek city and a certain amount of territory outside the city. Each city-state had its own unique political structure for its citizens: some were oligarchs where political power was maintained in the hands of a few individuals or aristocratic families; others were ruled by tyrants (or sole rulers) who sometimes took power by force, but who often had a lot of support from the public. These included people such as Solon, Peisistratus and Cleisthenes in Athens.

The reforms by Cleisthenes led to the introduction of the Athenian democracy. Power was placed in the hands of the citizens who were male (women or slaves did not participate). It was an extremely liberal democracy where citizens voted on all of the important issues. Often, this led to disastrous results as speakers who were skilled in rhetoric could exert significant influence. This led Plato to advocate rule by philosopher kings rather than democracy.

Early Greek mathematics commenced approximately 500–600 B.C. with work done by Pythagoras and Thales. Pythagoras was a philosopher and mathematician who had spent time in Egypt becoming familiar with Egyptian mathematics. He was born on the island of Samos, and he later moved to Croton in the south of Italy. He formed a secret society known as the Pythagoreans, and they included men and women who believed in the transmigration of souls and that number was the essence of all things. They discovered the mathematics for harmony in music, with the relationship between musical notes being expressed in numerical ratios of small whole numbers. Pythagoras is credited with the discovery of Pythagoras' theorem, although the Babylonians probably knew this theorem about 1000 years before Pythagoras. The Pythagorean society was dealt a major blow[13] by the discovery of

---

[12] The length of a side of the bottom base of the pyramid is b_1 and the length of a side of the top base is b_2.

[13] The Pythagoreans were a secret society and its members took a vow of silence with respect to this discovery. However, one member of the society is said to have shared the secret result with others outside the sect, and the apocryphal account is that he was thrown into a lake for his betrayal and drowned. They obviously took mathematics seriously back then!

the incommensurability of the square root of 2: i.e. there are no numbers p, q such that $\sqrt{2}=p/q$.

Thales was a sixth-century (B.C.) philosopher from Miletus in Asia Minor who made contributions to philosophy, geometry and astronomy. His contributions to philosophy are mainly in the area of metaphysics, and he was concerned with questions on the nature of the world. His objective was to give a natural or scientific explanation of the cosmos, rather than relying on the traditional supernatural explanation of creation in Greek mythology. He believed that there was a single substance that was the underlying constituent of the world, and he believed that this substance was water. He also contributed to mathematics [AnL:95], and Thales' theorem in Euclidean geometry states that if A, B and C are points on a circle, where the line AC is a diameter of the circle, then the angle $<ABC$ is a right angle.

The rise of Macedonia led to the Greek city-states being conquered by Philip of Macedonia in the fourth century B.C. His son, Alexander the Great, defeated the Persian Empire, and he extended his empire to include most of the known world. This led to the *Hellenistic Age* where Greek language and culture spread to the known world. Alexander founded the city of Alexandria, and it became a major centre of learning in Ptolemaic Egypt.[14] However, Alexander's reign was very short as he died at the young age of 33 in 323 B.C.

Euclid lived in Alexandria during the early Hellenistic period. He is considered the father of geometry and the deductive method in mathematics. His systematic treatment of geometry and number theory is published in *The Thirteen Books of The Elements* [Hea:56]. It starts from five axioms, five postulates and 23 definitions to logically derive a comprehensive set of theorems. His method of proof was generally constructive, in that as well as demonstrating the truth of a theorem, the proof would often include the construction of the required entity. It was also used as indirect proof (a nonconstructive proof) to show that there are an infinite number of primes:

1. Suppose there are a finite number of primes (say n primes).
2. Multiply all n primes together and add 1 to form N.
   $(N = p_1 * p_2 * \ldots * p_n + 1)$
3. N is not divisible by $p_1, p_2, \ldots, p_n$ as dividing by any of these gives a remainder of one.
4. Therefore, N must either be prime or divisible by some other prime that was not included in the list.
5. Therefore, there must be at least $n+1$ primes.
6. This is a contradiction (it was assumed that there are n primes).
7. Therefore, the assumption that there are a finite number of primes is false.
8. Therefore, there are an infinite number of primes.

[14] The ancient library in Alexandria was once the largest library in the world. It was build during the Hellenistic period in the third century B.C. and destroyed by fire in 391 A.D.

Euclidean geometry included the parallel postulate (the fifth postulate). This postulate generated interest, as many mathematicians believed that it was unnecessary and could be proved as a theorem. It states that:

Definition 2.1 (Parallel Postulate) *If a line segment intersects two straight lines forming two interior angles on the same side that sum to less than two right angles, then the two lines, if extended indefinitely, meet on that side on which the angles sum to less than two right angles.*

This postulate was later proved to be independent of the other postulates with the development of non-Euclidean geometries in the nineteenth century. These include the hyperbolic geometry discovered independently by Bolyai and Lobachevsky and elliptic geometry as developed by Riemann. The standard model of Riemannian geometry is the sphere where lines are great circles.

The material in the *Euclid's Elements* is a systematic development of geometry starting from the small set of axioms, postulates and definitions, leading to theorems derived logically from the axioms and postulates. There are some jumps in reasoning, and the German mathematician, David Hilbert, later added extra axioms to address this. Euclidean geometry contains many well-known mathematical results such as Pythagoras' theorem, Thales' theorem, sum of angles in a triangle, prime numbers, greatest common divisor and least common multiple, Euclidean algorithm, areas and volumes, tangents to a point and algebra.

The Euclidean algorithm is one of the oldest known algorithms, and it is used to determine the greatest common divisor of two numbers a and b. It is presented in *The Elements*, but it was known well before Euclid. The formulation of the *gcd* algorithm for two natural numbers a and b is as follows:

1. Check if b is zero. If so, then a is the gcd.
2. Otherwise, the gcd (a, b) is given by gcd $(b, a \bmod b)$.

It is also possible to determine integers p and q such that $ap+bq=\gcd(a, b)$.

The proof of the Euclidean algorithm is as follows. Suppose a and b are two positive numbers whose gcd is to be determined, and let r be the remainder when a is divided by b:

1. Clearly $a=qb+r$ where q is the quotient of the division.
2. Any common divisor of a and b is also a divider or r (since $r=a-qb$).
3. Similarly, any common divisor of b and r will also divide a.
4. Therefore, the greatest common divisor of a and b is the same as the greatest common divisor of b and r.
5. The number r is smaller than b and we will reach $r=0$ in finitely many steps.
6. The process continues until $r=0$.

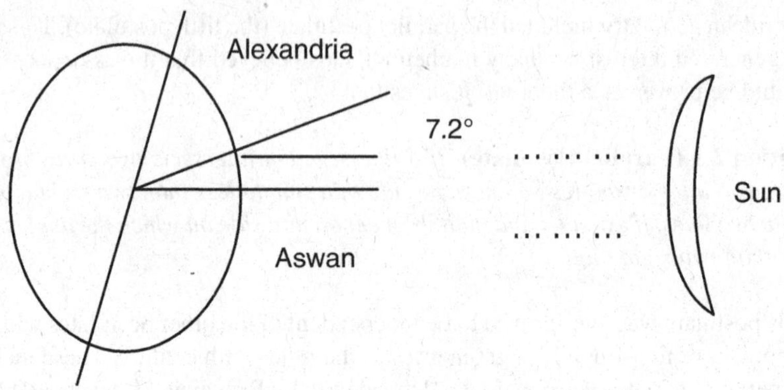

Fig. 2.6 Eratosthenes measurement of the circumference of the Earth

Comment 2.1 *Algorithms are fundamental in computing as they define the procedure by which a problem is solved. A computer program implements the algorithm in some programming language.*

Eratosthenes was a Hellenistic mathematician and scientist who worked as librarian in the famous library in Alexandria. He devised a system of latitude and longitude, and he became the first person to estimate of the size of the circumference of the Earth (Fig. 2.6). His approach to the calculation was as follows:

1. Eratosthenes believed that the Earth was a sphere.
2. On the summer solstice at noon in the town of Syene (ancient name of Aswan[15]) on the Tropic of Cancer in Egypt, the sun appears directly overhead.
3. He assumed that rays of light came from the sun in parallel beams and reached the Earth at the same time.
4. At the same time in Alexandria, he had measured that the sun would be 7.2° south of the zenith.
5. He assumed that Alexandria was directly north of Aswan.
6. He concluded that the distance from Alexandria to Aswan was $^{7.2}/_{360}$ of the circumference of the Earth.
7. The distance between Alexandria and Aswan was 5000 stadia (approximately 800 km).
8. He established a value of 252,000 stadia or approximately 396,000 km (the actual circumference at the equator is 40,075 km).

Eratosthenes' calculation was an impressive result for 200 B.C. The errors in his calculation were due to:

[15] The town of Aswan is famous today for the Aswan high dam, which was built in the 1960s. There was an older Aswan dam built by the British in the late nineteenth century. The new dam led to a rise in the water level of Lake Nasser and flooding of archaeological sites along the Nile. Several sites such as Abu Simbel and the island of Philae were relocated to higher ground.

1. Aswan is not exactly on the Tropic of Cancer but it is actually 55 km north of it.
2. Alexandria is not exactly north of Aswan and there is a difference of 3° longitude.
3. The distance between Aswan and Alexandria is 729 km not 800 km.
4. Angles in antiquity could not be measured with absolute precision.
5. The angular distance is actually 7.08 ° and not 7.2°.

Eratosthenes also calculated the approximate distance to the moon and sun, and he also produced maps of the known world. He developed a useful algorithm for determining all of the prime numbers up to a specified integer, and this is known as the *Sieve of Eratosthenes*. The steps in the algorithm are as follows:

1. Write a list of the numbers from 2 to the largest number to be tested. This first list is called A.
2. A second list B is created to list the primes. It is initially empty.
3. The number 2 is the first prime number, and it is added to list B.
4. Strike off (or remove) all multiples of 2 from list A.
5. The first remaining number in list A is a prime number and this prime number is added to list B.
6. Strike off (or remove) this number and all multiples of it from list A.
7. Repeat steps 5 through 7 until no more numbers are left in list A.

Comment 2.2 *The Sieve of Eratosthenes method is a well-known algorithm for determining prime numbers.*

Archimedes was a mathematician and astronomer who lived in Syracuse, Sicily. He discovered the law of buoyancy known as Archimedes' principle.

The buoyancy force is equal to the weight of the displaced fluid.

He is believed to have discovered the principle while sitting in his bath, and he was so overwhelmed with his discovery that he rushed out onto the streets of Syracuse shouting *Eureka*, forgetting to put on his clothes.

The weight of the displaced liquid is proportional to the volume of the displaced liquid. Therefore, if two objects have the same mass, the one with greater volume (or smaller density) has greater buoyancy. An object will float if its buoyancy force (i.e. the weight of liquid displaced) exceeds the downward force of gravity (i.e. its weight). If the object has exactly the same density as the liquid, then it will stay still, neither sinking nor floating upwards.

For example, a rock is generally a very dense material, and so it usually does not displace its own weight. Therefore, a rock will sink to the bottom as the downward weight exceeds the buoyancy weight. However, the weight of a buoyancy device is significantly less than the liquid that it would displace, and so it floats at a level where it displaces the same weight of liquid as the weight of the object.

Archimedes also made good contributions to mathematics including an approximation to π, contributions to the positional numbering system, geometric series, mathematics and physics. He also solved several interesting problems: e.g. the calculation of the composition of cattle in the herd of the sun god by solving a number of simultaneous *Diophantine equations* (named after Diophantus). The herd consisted of bulls and cows, with one part of the herd consisting of white, second part black, third spotted and the fourth brown. Various constraints were then expressed in Diophantine equations, and the problem was to determine the precise composition of the herd.

He calculated the number of grains of sands in the known universe and challenged the prevailing view this was too large to be determined. He developed a naming system for large numbers, as the largest number in use at the time was a myriad (100 million), where a myriad is 10,000. He developed the laws of exponents, i.e. $10^a 10^b = 10^{a+b}$, and his calculation of the upper bound includes not only the grains of sand on each beach but on the Earth filled with sand and the known universe filled with sand. His final estimate of the upper bound of the number of grains of sand in a filled universe was 10^{64}.

It is possible that he may have developed the odometer,[16] which could calculate the total distance travelled on a journey. An odometer is described by the Roman engineer Vitruvius around 25 B.C. It employed a wheel with a diameter of 4 ft, and the wheel turned 400 times in every mile.[17] The device included gears and pebbles and a 400 tooth cogwheel that turned once every mile and caused one pebble to drop into a box. The total distance travelled was determined by counting the number of pebbles in the box.

Aristotle was born in Macedonia and he became a student of Plato at Plato's academy in Athens in the fourth century B.C. (Fig. 2.7). Aristotle later founded his own school (known as the Lyceum) in Athens, and he was also the tutor of Alexander the Great. He made contributions to biology, logic, politics, ethics and metaphysics.

His starting point to knowledge acquisition was the senses, as he believed that these were essential to acquire knowledge. His position is the opposite of Plato who argued that the senses deceive and should not be relied upon. Plato's writings are mainly in dialogues involving his former mentor Socrates.[18]

Aristotle made important contributions to formal reasoning with his development of syllogistic logic. Syllogistic logic (also known as term logic) consists of

---

[16] The origin of the word 'odometer' is from the Greek words 'οδος' (meaning journey) and μετρον (meaning measure).

[17] The figures given here are for the distance of one Roman mile. This is given by $\pi\, 4 * 400 = 12.56 * 400 = 5024$ (which is less than 5280 ft for a standard mile in the imperial system).

[18] Socrates was a moral philosopher who deeply influenced Plato. His method of enquiry into philosophical problems and ethics was by questioning. Socrates himself maintained that he knew nothing (Socratic ignorance). However, from his questioning it became apparent that those who thought they were clever were not really that clever after all. His approach obviously would not have made him very popular with the citizens of Athens. Socrates had consulted the oracle at Delphi to find out who was the wisest of all men, and he was informed that there was no one wiser than him. Socrates was sentenced to death for allegedly corrupting the youth of Athens, and he was forced to drink the juice of the hemlock plant (a type of poison).

Fig. 2.7 Plato and Aristotle

reasoning with two premises and one conclusion. Each premise consists of two terms and there is a common middle term. The conclusion links the two unrelated terms from the premises. For example,

Premise1	All Greeks are mortal.
Premise2	Socrates is a Greek.
Conclusion	Socrates is mortal.

The common middle term is 'Greek', which appears in the two premises. The two unrelated terms from the premises are 'Socrates' and 'mortal'. The relationship

Table 2.1 Syllogisms: relationship between terms

Relationship	Abbr.
Universal affirmation	A
Universal negation	E
Particular affirmation	I
Particular negation	O

between the terms in the first premise is that of the universal: i.e. anything or any person that is a Greek is mortal. The relationship between the terms in the second premise is that of the particular: i.e. Socrates is a person that is a Greek. The conclusion from the two premises is that Socrates is mortal: i.e. a particular relationship between the two unrelated terms 'Socrates' and 'mortal'.

The example above is an example of a valid syllogistic argument. Aristotle studied the various possible syllogistic arguments and determined those that were valid and those that were invalid. There are several candidate relationships that may potentially exist between the terms in a premise. These are listed in Table 2.1.

In general, a syllogistic argument will be of the form

$$\begin{array}{c} \mathrm{S}x\mathrm{M} \\ \mathrm{M}y\mathrm{P} \\ \hline \mathrm{S}z\mathrm{P} \end{array}$$

where x, y, z may be universal affirmation, universal negation, particular affirmation and particular negation. Syllogistic logic is described in more detail in [ORg:12]. Aristotle's work was highly regarded in classical and mediaeval times, and Kant believed that there was nothing else to invent in logic.

An early form of propositional logic that was developed by Chrysippus[19] in the third-century B.C. Aristotelian logic is of historical interest today, and it has been replaced by propositional and predicate logic.

2.5 The Romans

Rome is said to have been founded[20] by Romulus and Remus about 750 B.C. Early Rome covered a small part of Italy, but it gradually expanded in size and importance. It destroyed Carthage[21] in 146 B.C. to become the major power in the

[19] Chrysippus was the head of the Stoics in the third century B.C.

[20] The Aeneid by Virgil suggests that the Romans were descended from survivors of the Trojan War and that Aeneas brought surviving Trojans to Rome after the fall of Troy.

[21] Carthage was located in Tunisia, and the wars between Rome and Carthage are known as the Punic wars. Hannibal was one of the great Carthaginian military commanders, and during the second Punic war, he brought his army to Spain, marched through Sprain and crossed the Pyrenees. He then marched along southern France and crossed the Alps into Northern Italy. His army also consisted of war elephants. Rome finally defeated Carthage and levelled the city.

Fig. 2.8 Julius Caesar

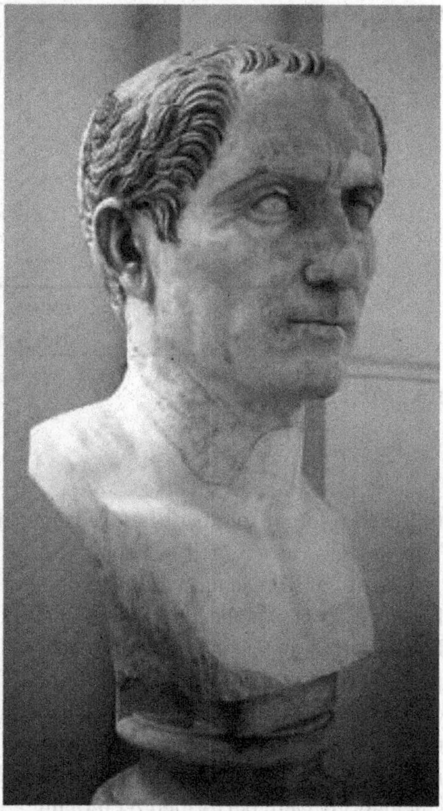

Mediterranean. The Romans colonized the Hellenistic world, and they were influenced by Greek culture and mathematics. Julius Caesar (Fig. 2.8) conquered the Gauls in 58 B.C.

The Gauls consisted of several disunited Celtic[22] tribes. Vercingetorix succeeded in uniting them, but he was defeated at the siege of Alesia in 52 B.C.

The Roman number system uses letters to represent numbers (Fig. 2.9) and a number consists of a sequence of letters. The evaluation rules specify that if a large number follows a smaller number, then the smaller number is subtracted from the large: e.g. IX represents 9 and XL represents 40. Similarly, if a smaller number

[22] The Celtic period commenced around 1000 B.C. in Hallstatt (near Salzburg in Austria). The Celts were skilled in working with iron and bronze, and they gradually expanded into Europe. They eventually reached Britain and Ireland around 600 B.C. The early Celtic period was known as the 'Hallstatt period', and the later Celtic period is known as the 'La Téne' period. The La Téne period is characterized by the quality of ornamentation produced. The Celtic museum in Hallein in Austria provides valuable information and artefacts on the Celtic period. The Celtic language has similarities to the Irish language. However, the Celts did not employ writing, and the ogham writing developed in Ireland was developed in the early Christian period.

Fig. 2.9 Roman numbers

I	= 1
V	= 5
X	= 10
L	= 50
C	= 100
D	= 500
M	= 1000

Alphabet Symbol	abcde fghij klmno pqrst uvwxyz
Cipher Symbol	dfegh ijklm nopqr stuvw xyzabc

Fig. 2.10 Caesar cipher

followed a larger number, they were generally added: e.g. MCC represents 1200. They had no zero in their system.

The use of Roman numerals was cumbersome, and an abacus was often employed for calculation. An abacus consists of several columns in which pebbles are placed. Each column represented powers of 10: i.e. 10^0, 10^1, 10^2, 10^3, etc. The column to the far right represents 1; the column to the left 10; next column to the left 100; and so on. Pebbles (calculi) were placed in the columns to represent different numbers: e.g. the number represented by an abacus with four pebbles on the far right, two pebbles in the column to the left and three pebbles in the next column to the left is 324. The calculations were performed by moving pebbles from column to column.

Merchants introduced a set of weights and measures (including the *libra* for weights and the *pes* for lengths). They developed an early banking system to provide loans for business and commenced minting coins around 290 B.C. The Romans also made contributions to calendars, and Julius Caesar introduced the Julian calendar in 45 B.C. It has a regular year of 365 days divided into 12 months and a leap day is added to February every 4 years. However, too many leap years are added over time, and this led to the introduction of the Gregorian calendar in 1582.

Caesar employed a *substitution cipher* (Fig. 2.10) on his military campaigns to ensure that important messages were communicated safely. This involves the substitution of each letter in the *plaintext* (i.e. the original message) by a letter with a fixed number of positions down in the alphabet. For example, a shift of three positions causes the letter B to be replaced by E, the letter C by F and so on. The Caesar cipher is easily broken, as the frequency distribution of letters may be employed to determine the mapping. The cipher is defined as follows.

The process of enciphering a message (i.e. plaintext) involves mapping each letter in the plaintext to the corresponding cipher letter. For example, the encryption of 'summer solstice' involves

Plaintext :	Summer Solstice
Cipher Text	vxpphu vrovwleh

The decryption involves the reverse operation: i.e. for each cipher letter, the corresponding plaintext letter is identified from the table:

Cipher Text	vxpphu vrovwleh
Plaintext :	Summer Solstice

The encryption may also be done using modular arithmetic. The numbers 0–25 represent the alphabet letters, and addition (modula 26) is used to perform the encryption. The encoding of the plaintext letter x is given by

$$c = x + 3 \pmod{26}$$

Similarly, the decoding of a cipher letter represented by the number c is given by

$$x = c - 3 \pmod{26}$$

The emperor Augustus[23] employed a similar substitution cipher (with a shift key of 1). The Caesar cipher remained in use up to the early twentieth century. However, by then frequency analysis techniques were available to break the cipher. The Vigenère cipher uses a Caesar cipher with a different shift at each position in the text. The value of the shift to be employed with each plaintext letter is defined using a repeating keyword.

2.6 Islamic Influence

Islamic mathematics refers to mathematics developed in the Islamic world from the birth of Islam in the early seventh century up until the seventeenth century. The Islamic world commenced with the prophet Mohammed in Mecca and spread throughout the Middle East, North Africa and Spain. The golden age of Islamic civilization was from 750 A.D. to 1250 A.D., and during this period enlightened caliphs recognized the value of knowledge and sponsored scholars to come to Baghdad to gather and translate the existing world knowledge into Arabic.

This led to the preservation of the Greek texts during the dark ages in Europe. Further, the Islamic cities of Baghdad, Cordoba and Cairo became key intellectual

[23] Augustus was the first Roman emperor, and his reign ushered in a period of peace and stability following the bitter civil wars. He was the adopted son of Julius Caesar and was called Octavion before he became emperor. The earlier civil wars were between Caesar and Pompey, and following Caesar's assassination civil war broke out between Mark Anthony and Octavion. Octavion defeated Anthony and Cleopatra at the battle of Actium in 31 B.C.

Fig. 2.11 Mohammed Al-Khwarizmi

centres, and scholars added to existing knowledge (e.g. in mathematics, astronomy, medicine and philosophy), as well as translating the known knowledge into Arabic.

The Islamic mathematicians and scholars were based in several countries in the Middle East, North Africa and Spain. Early work commenced in Baghdad, and the mathematicians were also influenced by the work of Hindu mathematicians, who had introduced the decimal system and decimal numerals. Among the well-known Islamic scholars are Ibn Al-Haytham, a tenth-century Iraqi scientist; Mohammed Al-Khwarizmi (Fig. 2.11), a ninth Persian mathematician; Abd Al-Rahman al-Sufi, a Persian astronomer who discovered the Andromeda galaxy; Ibn Al-Nafis, a Syrian who did work on circulation in medicine; Averroes, who was an Aristotelian philosopher from Cordoba in Spain; Avicenna who was a Persian philosopher; and Omar Khayyam who was a Persian mathematician and poet.

Many caliphs (Muslim rulers) were enlightened and encouraged scholarship in mathematics and science. They set up a centre for translation and research in Baghdad, and existing Greek texts such as the works of Euclid, Archimedes, Apollonius and Diophantus were translated into Arabic. Al-Khwarizmi made contributions to early classical algebra, and the word algebra comes from the Arabic word *al jabr* that appears in a textbook by Al-Khwarizmi. The origin of the word *algorithm* is from the name of the Islamic scholar 'Al-Khwarizmi'.

Education was important during the golden age, and the Al-Azhar University in Cairo (Fig. 2.12) was established in 970 A.D., and the Al-Qarawiyyin University in Fez, Morocco, was established in 859 A.D. The Islamic world has created beautiful architecture and art including the ninth-century Great Mosque of Samarra in Iraq, the tenth-century Great Mosque of Cordoba and the eleventh-century Alhambra in Grenada.

Fig. 2.12 Al-Azhar University, Cairo

The Moors[24] invaded Spain in the eighth century A.D., and they ruled large parts of the Peninsula for several centuries. Moorish Spain became a centre of learning, and this led to Islamic and other scholars coming to study at the universities in Spain. Many texts on Islamic mathematics were translated from Arabic into Latin, and these were invaluable in the renaissance in European learning and mathematics from the thirteenth century. The Moorish influence[25] in Spain continued until the time of the Catholic Monarchs[26] in the fifteenth century. Ferdinand and Isabella united Spain, defeated the Moors in Andalusia and expelled them from Spain.

[24] The origin of the word 'Moor' is from the Greek work μυορος meaning very dark. It referred to the fact that many of the original Moors who came to Spain were from Egypt, Tunisia and other parts of North Africa.

[25] The Moorish influence includes the construction of various castles (*alcazar*), fortresses (*alcalzaba*) and mosques. One of the most striking Islamic sites in Spain is the palace of Alhambra in Granada, and it represents the zenith of Islamic art.

[26] The Catholic Monarchs refer to Ferdinand of Aragon and Isabella of Castille who married in 1469. They captured Granada (the last remaining part of Spain controlled by the Moors) in 1492.

The Islamic contribution to algebra was an advance on the achievements of the Greeks. They developed a broader theory that treated rational and irrational numbers as algebraic objects and moved away from the Greek concept of mathematics as being essentially geometry. Later Islamic scholars applied algebra to arithmetic and geometry and studied curves using equations. This included contributions to reduce geometric problems such as duplicating the cube to algebraic problems. Eventually this led to the use of symbols in the fifteenth century such as

$$x^n.x^m = x^{m+n}.$$

The poet Omar Khayyam was also a mathematician who did work on the classification of cubic equations with geometric solutions. Other scholars made contributions to the theory of numbers: e.g. a theorem that allows pairs of amicable numbers to be found. Amicable numbers are two numbers such that each is the sum of the proper divisors of the other. They were aware of Wilson's theory in number theory: i.e. for p prime, then p divides $(p-1)! + 1$.

The Islamic world was tolerant of other religious belief systems during the golden age, and there was freedom of expression provided that it did not infringe on the rights of others. It began to come to an end following the Mongol invasion and sack of Baghdad in the late 1250s and the Crusades. It continued to some extent until the conquest by Ferdinand and Isabella of Andalusia in the late fifteenth century.

2.7 Chinese and Indian Mathematics

The development of mathematics commenced in China about 1000 B.C., and it was independent of developments in other countries. The emphasis was on problem solving rather than on conducting formal proofs. It was concerned with finding the solution to practical problems such as the calendar, the prediction of the positions of the heavenly bodies, land measurement, conducting trade and the calculation of taxes.

The Chinese employed counting boards as mechanical aids for calculation from the fourth century B.C. Counting boards are similar to abaci and are usually made of wood or metal and contained carved grooves between which beads, pebbles or metal discs were moved. The abacus is a device, usually of wood having a frame that holds rods with freely sliding beads mounted on them. It is used as a tool to assist calculation, and it is useful for keeping track of the sums, the carry and so on of calculations.

Early Chinese mathematics was written on bamboo strips and included work on arithmetic and astronomy. The Chinese method of learning and calculation in mathematics was *learning by analogy*. This involves a person acquiring knowledge from observation of how a problem is solved and then applying this knowledge for problem solving to similar kinds of problems.

They had their version of Pythagoras' theorem and applied it to practical problems. They were familiar with the Chinese remainder theorem, the formula for finding the area of a triangle, as well as showing how polynomial equations (up to degree 10), could be solved. Other Chinese mathematicians showed how geometric problems could be solved by algebra, how roots of polynomials could be solved, how quadratic and simultaneous equations could be solved and how the area of various geometric shapes such as rectangles, trapezia and circles could be computed. Chinese mathematicians were familiar with the formula to calculate the volume of a sphere. The best approximation that the Chinese had of π was 3.14159, and this was obtained by approximations from inscribing regular polygons with $3 \times 2n$ sides in a circle.

The Chinese made contributions to number theory including the summation of arithmetic series and solving simultaneous congruences. The Chinese remainder theorem deals with finding the solutions to a set of simultaneous congruences in modular arithmetic. Chinese astronomers made accurate observations, which were used to produce a new calendar in the sixth century. This was known as the Taming calendar and it was based on a cycle of 391 years.

Indian mathematicians have made important contributions such as the development of the decimal notation for numbers that is now used throughout the world. This was developed in India sometime between 400 B.C. and 400 A.D. Indian mathematicians also invented zero and negative numbers and also did early work on the trigonometric functions of sine and cosine. The knowledge of the decimal numerals reached Europe through Arabic mathematicians, and the resulting system is known as the *Hindu-Arabic numeral system*.

The Sulva Sutras is a Hindu text that documents Indian mathematics and it dates from about 400 B.C. The Indians were familiar with the statement and proof of Pythagoras' theorem, rational numbers, quadratic equations as well as the calculation of the square root of 2 to five decimal places

2.8 Review Questions

1. Discuss the strengths and weaknesses of the various number systems.
2. Describe ciphers used during the Roman civilization and write a program to implement one of these.
3. Discuss the nature of an algorithm and its importance in computing.
4. Discuss the working of an abacus and its application to calculation.
5. What are the differences between syllogistic logic and propositional and predicate logic??

2.9 Summary

The last decades of the twentieth century have witnessed a proliferation of high-tech computers, mobile phones and information technology. Software is now pervasive and is in automobiles, airplanes, televisions and mobile communication. It is only in recent decades that technology has become an integral part of the western world, and the pace of change has been extraordinary. It has led to increases in industrial productivity and potentially allows humans the freedom to engage in more creative and rewarding tasks.

This chapter considered the contributions of early civilizations to computing and included a discussion on the Babylonians, the Egyptians, the Greeks and the Romans and Islamic scholars.

The Babylonian civilization flourished from about 2000 B.C., and they produced clay cuneiform tablets containing mathematical texts. These included tables for multiplication, division, squares and square roots, as well as the calculation of area and the solution of linear and quadratic equations.

The Egyptian civilization developed along the Nile from about 4000 B.C., and they used mathematics for practical problem solving such as measuring the annual Nile flooding and their building programme.

The Greeks made major contributions to western civilization. Euclid developed a systematic treatment of geometry starting from a small set of axioms, postulates and definitions to derive and prove a comprehensive set of theorems. Aristotle's syllogistic logic remained in use until the development of propositional and predicate logic in the late nineteenth century.

The Islamic contribution helped to preserve western knowledge during the dark ages in Europe. Islamic scholars in Baghdad, Cairo and Cordoba translated Greek texts into Arabic. They also added to existing knowledge in mathematics, science, astronomy and medicine.

Foundations of Computing 3

Abstract
This chapter discusses the foundations of computing, including the binary number system and the step reckoner calculating machine, which were invented by Leibniz. The difference engine was designed by Babbage to evaluate polynomials and to produce accurate mathematical tables. Babbage's design of the analytic engine provided the vision of a modern computer, and he envisaged that it would be analogous to the operation of the *Jacquard loom*, which is designed to weave (i.e. execute on the loom) a design pattern represented by a set of cards. Boole's symbolic logic provides the foundation for digital computing.

Key Topics
Leibniz
Binary numbers
Step reckoner
Babbage
Difference engine
Analytic engine
Lovelace
Boole
Shannon
Switching circuits

G. O'Regan, *Introduction to the History of Computing*, Undergraduate Topics in Computer Science, DOI 10.1007/978-3-319-33138-6_3

3.1 Introduction

This chapter considers important foundational work done by Wilhelm Leibniz, Charles Babbage, George Boole, Ada Lovelace and Claude Shannon. Leibniz was a seventeenth-century German mathematician, philosopher and inventor, and he is recognized (with Isaac Newton) as the inventor of calculus. He developed a calculating machine that could perform all of the four basic arithmetic operations (i.e. addition, subtraction, multiplication and division), and he also invented the binary number system, which is used extensively in the computer field.

Boole and Babbage are considered grandfathers of the computing field, with Babbage's analytic engine providing a vision of a mechanical computer and Boole's logic providing the foundation for modern digital computers.

Charles Babbage was a nineteenth-century scientist and inventor who did pioneering work on calculating machines. He invented the difference engine (a sophisticated calculator that could be used for the production of mathematical tables), and he also designed the analytic engine (the world's first mechanical computer). The design of the analytic engine included a processor, a memory and a way to input information and output results.

Lady Ada Lovelace was introduced into Babbage's ideas on the analytic engine at a dinner party. She was fascinated and predicted that such a machine could be used to compose music, produce graphics as well as solve mathematical and scientific problems. She explained how the analytic engine could be programmed, and she wrote what is considered the first computer program.

Boole was a nineteenth-century English mathematician who made important contributions to mathematics, probability theory and logic. Boole's logic provides the foundation for digital computers.

Claude Shannon was the first person to see the applicability of Boole's logic to switching theory, and it is the foundation for all modern digital computers. Shannon was a twentieth-century American mathematician and engineer who showed that Boolean algebra could simplify the design of circuits and telephone routing switches and that it provided the perfect mathematical model for switching theory and for the subsequent design of digital circuits and computers.

3.2 Step Reckoner Calculating Machine

Wilhelm Gottfried Leibniz (Fig. 3.1) was a German philosopher, mathematician and inventor in the field of mechanical calculators. He developed the binary number system used in digital computers, and he invented the calculus independently of Sir Isaac Newton. He became familiar with Pascal's calculating machine, the *Pascaline*, while in Paris in the early 1670s. He recognized its limitations as the machine could perform addition and subtraction operations only.

He designed and developed a calculating machine that could perform addition, subtraction, multiplication, division and the extraction of roots. He commenced work on the machine in 1672, and the machine was completed in 1694. It was the

Fig. 3.1 Wilhelm Gottfried Leibniz

first calculator that could perform all four arithmetic operations, and it was superior to the existing Pascaline machine. Leibniz's machine was called the *step reckoner* (Fig. 3.2), and it allowed the common arithmetic operations to be carried out mechanically.

The operating mechanism used in his calculating machine was based on a counting device called the stepped cylinder or *Leibniz wheel*. This mechanism allowed a gear to represent a single decimal digit from 0 to 9 in just one revolution, and this mechanism remained the dominant approach in the design of calculating machines for the next 200 years. The Leibniz wheel was essentially a counting device consisting of a set of wheels that were used in calculation. The step reckoner consisted of an accumulator which could hold 16 decimal digits and an 8-digit input section. The eight dials at the front of the machine set the operand number, which was then employed in the calculation.

The machine performed multiplication by repeated addition and division by repeated subtraction. The basic operation is to add or subtract the operand from the accumulator as many times as desired. The machine could add or subtract an 8-digit number to the 16-digit accumulator to form a 16-digit result. It could multiply two 8-digit numbers to give a 16-digit result, and it could divide a 16-bit number by an 8-digit number. Addition and subtraction are performed in a single step, with the

Fig. 3.2 Replica of step reckoner at Technische Sammlungen Museum, Dresden

operating crank turned in the opposite direction for subtraction. The result is stored in the accumulator.

3.3 Binary Numbers

Arithmetic has traditionally been done using the decimal notation,[1] and Leibniz was one of the first to recognize the potential of the binary number system. This system uses just two digits, namely, '0' and '1', with the number 2 represented by 10, the number 4 by 100 and so on. Leibniz described the binary system in *Explication de l'Arithmétique Binaire* [Lei:03], which was published in 1703. A table of values for the first 15 binary numbers is given in Table 3.1.

Leibniz's 1703 paper describes how binary numbers may be added, subtracted, multiplied and divided, and he was an advocate of their use. The key advantage of the use of binary notation is in digital computers, where a binary digit may be implemented by an *on-off switch*, with the digit 1 representing that the switch is on and the digit 0 representing that the switch is off.

[1]The segadecimal (or base 60) system was employed by the Babylonians (as discussed in Chapter 2). The decimal system was developed by Indian and Arabic mathematicians between 800 and 900 A.D., and it was introduced to Europe in the late twelfth/early thirteenth century. It is known as the *Hindu-Arabic system*.

Table 3.1 Binary number system

Binary	Dec.	Binary	Dec.	Binary	Dec.	Binary	Dec.
0000	0	0100	4	1000	8	1100	12
0001	1	0101	5	1001	9	1101	13
0010	2	0110	6	1010	10	1110	14
0011	3	0111	7	1011	11	1111	15

The use of binary arithmetic allows more complex mathematical operations to be performed by relay circuits, and Boolean logic (described in a later section) is the perfect model for simplifying such circuits and is the foundation underlying digital computing.

3.4 The Difference Engine

Charles Babbage (Fig. 3.3) is considered (along with George Boole) to be one of the grandfathers of the computing field. He made contributions to several areas including mathematics, statistics, astronomy, calculating machines, philosophy, railways and lighthouses. He founded the British Statistical Society and the Royal Astronomical Society.

Babbage was interested in accurate mathematical tables as these are essential for navigation and scientific work. However, there was a high error rate in the existing tables due to human error introduced during calculation. He became interested in the problem of finding a mechanical method to perform the calculations, as this would eliminate errors introduced by human calculation. Babbage wished to develop a more advanced machine than Pascal's Pascaline or Leibniz's step reckoner, which were limited to the basic arithmetic operations. He wished to develop a machine that could compute polynomial functions.

He designed the difference engine (no. 1) in 1821 for the production of mathematical tables. This is essentially a mechanical calculator (analogous to modern electronic calculators), and it was designed to compute polynomial functions of degree 4. It could also compute logarithmic and trigonometric functions such as sine or cosine (as these may be approximated by polynomials).[2]

The accurate approximation of trigonometric, exponential and logarithmic functions by polynomials depends on the degree of the polynomials, on the number of decimal digits that it is being approximated to and on the error function. A higher-degree polynomial is generally able to approximate the function more accurately.

---

[2]The power series expansion of the sine function is given by $\text{Sin}(x) = x - x^3/3! + x^5/5! - x^7/7! + \ldots$. The power series expansion for the cosine function is given by $\text{Cos}(x) = 1 - x^2/2! + x^4/4! - x^6/6! + \ldots$. Functions may be approximated by interpolation, and the approximation of a function by a polynomial of degree n requires $n+1$ points on the curve for the interpolation. That is, the curve formed by the polynomial of degree n that passes through the $n+1$ points of the function to be approximated is an approximation to the function. The error function also needs to be considered.

Fig. 3.3 Charles Babbage

Babbage produced prototypes for parts of the difference engine, but he never actually completed the machine. The Swedish engineers, George and Edward Schuetz, built the first working difference engine (based on Babbage's design) in 1853 with funding from the Swedish government. Their machine could compute polynomials of degree 4 on 15-digit numbers, and the third Schuetz difference engine is on display at the Science Museum in London.

It was the first machine to compute and print mathematical tables mechanically. The machine was accurate, and it showed the potential of mechanical machines as a tool for scientists and engineers.

The machine is unable to perform multiplication or division directly. Once the initial value of the polynomial and its derivative are calculated for some value of x, the difference engine may calculate any number of nearby values using the numerical method of finite differences. This method replaces computational intensive tasks involving multiplication or division, by an equivalent computation that just involves addition or subtraction. The method of finite differences is described in more detail in [ORg:12].

The British government cancelled Babbage's project in 1842. He designed an improved difference engine no.2 (Fig. 3.4) in 1849. It could operate on seventh-order differences (i.e. polynomials of order 7) and 31-digit numbers. The machine consisted of eight columns with each column consisting of 31 wheels. However, it was over 150 years later before it was built (in 1991) to mark the 200th anniversary of his birth. The Science Museum in London also built the printer that Babbage

Fig. 3.4 Difference engine no. 2 (Photo public domain)

Table 3.2 Analytic engine

Part	Function
Store	This contains the variables to be operated upon as well as all those quantities, which have arisen from the result of intermediate operations
Mill	The mill is essentially the processor of the machine into which the quantities about to be operated upon are brought

designed, and both the machine and the printer worked correctly according to Babbage's design (after a little debugging).

3.5 The Analytic Engine: Vision of a Computer

The difference engine was designed to produce mathematical tables, and it required human intervention to perform the calculation. Babbage recognized its limitations, and he proposed a revolutionary solution by outlining his vision of a mechanical computer. His plan was to construct a new machine that would be capable of executing all tasks that may be expressed in algebraic notation. His vision of a computer (analytic engine) consisted of two parts (Table 3.2).

Babbage intended that the operation of the analytic engine would be analogous to the operation of the *Jacquard loom.*[3] The latter is capable of weaving (i.e. executing on the loom) a design pattern that has been prepared by a team of skilled artists. The design pattern is represented by a set of cards with punched holes, where each card represents a row in the design. The cards are then ordered and placed in the loom, and the loom produces the exact pattern.

The use of the punched cards in the analytic engine allowed the formulae to be manipulated in a manner dictated by the programmer. The cards commanded the analytic engine to perform various operations and to return a result. Babbage distinguished between two types of punched cards:

- *Operation cards*
- *Variable cards*

Operation cards are used to define the operations to be performed, whereas the variable cards define the variables or data that the operations are performed upon. His planned use of punched cards to store programs in the analytic engine is similar to the idea of a stored computer program in von Neumann architecture. However, Babbage's idea of using punched cards to represent machine instructions and data was over 100 years before digital computers. *Babbage's analytic engine is therefore an important milestone in the history of computing*.

Babbage intended that the program be stored on read-only memory using punched cards and that the input and output would be carried out using punched cards. He intended that the machine would be able to store numbers and intermediate results in memory that could then be processed. There would be several punch card readers in the machine for programs and data. He envisioned that the machine would be able to perform conditional jumps as well as parallel processing where several calculations could be performed at once.

The analytic engine was designed in 1834 as the world's first mechanical computer [Bab:42]. It included a processor, a memory and a way to input information and output results. However, the machine was never built, as Babbage was unable to secure funding from the British government.

[1]3 The Jacquard loom was invented by Joseph Jacquard in 1801. It is a mechanical loom which used the holes in punched cards to control the weaving of patterns in a fabric. The use of punched cards allowed complex designs to be woven from the pattern defined on the punched cards. Each punched card corresponds to one row of the design, and the cards were appropriately ordered. It was very easy to change the pattern of the fabric being weaved on the loom, as this simply involved changing cards.

Fig. 3.5 Lady Ada Lovelace

3.5.1 Applications of Analytic Engine

Lady Augusta Ada Lovelace (nee Byron)[4] (Fig. 3.5) was a mathematician who collaborated with Babbage on applications for the analytic engine. She is considered the world's first programmer, and the Ada programming language is named in her honour.

She was introduced to Babbage at a dinner party in 1833, and she visited Babbage's studio in London, where the prototype difference engine was on display. She recognized the beauty of its invention, and she was fascinated by the idea of the analytic engine. She communicated regularly with Babbage with ideas on its applications.

Lovelace produced an annotated translation of Menabrea's *Notions sur la machine analytique de Charles Babbage* [Lov:42]. She added copious notes to the translation,[5] which were about three times the length of the original memoir, and considered many of the difficult and abstract questions connected with the subject. These notes are regarded as a description of a computer and software.

---

[4] Lady Ada Lovelace was the daughter of the poet Lord Byron.

[5] There is some controversy as to whether this was entirely her own work or a joint effort by Lovelace and Babbage.

She explained in the notes how the analytic engine could be programmed and wrote what is considered to be the first computer program. This program detailed a plan be written for how the engine would calculate *Bernoulli numbers*. Lady Ada Lovelace is therefore considered to be the first computer programmer, and Babbage called her the 'enchantress of numbers'.

She saw the potential of the analytic engine to fields other than mathematics. She predicted that the machine could be used to compose music, produce graphics as well as solve mathematical and scientific problems. She speculated that the machine might act on other things apart from numbers and be able to manipulate symbols according to rules. In this way, a number could represent an entity other than a quantity.

3.6 Boole's Symbolic Logic

George Boole (Fig. 3.6) was born in Lincoln, England, in 1815. His father (a cobbler who was interested in mathematics and optical instruments) taught him mathematics and showed him how to make optical instruments. Boole inherited his father's interest in knowledge, and he was self-taught in mathematics and Greek. He taught at various schools near Lincoln, and he developed his mathematical knowledge by working his way through Newton's *Principia*, as well as applying himself to the work of mathematicians such as Laplace and Lagrange.

He published regular papers from his early twenties, and these included contributions to probability theory, differential equations and finite differences. He developed Boolean algebra, which is the foundation for modern computing, and he is

Fig. 3.6 George Boole

considered (along with Babbage) to be one of the grandfathers of computing. His work was theoretical, and he never actually built a computer or calculating machine. *However, Boole's symbolic logic was the perfect mathematical model for switching theory and for the design of digital circuits.*

Boole became interested in formulating a calculus of reasoning, and he published *The Mathematical Analysis of Logic* in 1847 [Boo:48]. This work developed novel ideas on a logical method, and he argued that logic should be considered as a separate branch of mathematics, rather than a part of philosophy. He argued that there are mathematical laws to express the operation of reasoning in the human mind, and he showed how Aristotle's syllogistic logic could be reduced to a set of algebraic equations. He corresponded regularly on logic with Augustus De Morgan.[6]

His paper on logic introduced two quantities '0 and '1'. He used the quantity 1 to represent the universe of thinkable objects (i.e. the universal set), and the quantity 0 represents the absence of any objects (i.e. the empty set). He then employed symbols such as x, y, z, etc., to represent collections or classes of objects given by the meaning attached to adjectives and nouns. Next, he introduced three operators (+, − and ×) that combined classes of objects.

The expression xy (i.e. x multiplied by y or $x \times y$) combines the two classes x, y to form the new class xy (i.e. the class whose objects satisfy the two meanings represented by class x *and* class y). Similarly, the expression $x+y$ combines the two classes x, y to form the new class $x+y$ (that satisfies either the meaning represented by class x *or* class y). The expression $x-y$ combines the two classes x, y to form the new class $x-y$. This represents the class that satisfies the meaning represented by class x but not class y. The expression $(1-x)$ represents objects that do not have the attribute that represents class x.

Thus, if x=black and y=sheep, then xy represents the class of black sheep. Similarly, $(1-x)$ would represents the class obtained by the operation of selecting all things in the world except black things; $x\,(1-y)$ represents the class of all things that are black but not sheep; and $(1-x)\,(1-y)$ would give us all things that are neither sheep nor black.

He showed that these symbols obeyed a rich collection of algebraic laws and could be added, multiplied, etc., in a manner that is similar to real numbers. These symbols may be used to reduce propositions to equations, and algebraic rules may be employed to solve the equations. The rules include:

1.	$x+0=x$	(Additive Identity)
2.	$x+(y+z)=(x+y)+z$	(Associative)
3.	$x+y=y+x$	(Commutative)
4.	$x+(1-x)=1$	
5.	$x\,1=x$	(Multiplicative Identity)
6.	$x\,0=0$	

[6] De Morgan was a nineteenth-century British mathematician based at University College London. De Morgan's laws in set theory and logic state that $(A \cup B)^c = A^c \cap B^c$ and $\neg (A \vee B) \equiv \neg A \wedge \neg B$.

7.	$x+1=1$	
8.	$xy=yx$	(Commutative)
9.	$x(yz)=(xy)z$	(Associative)
10.	$x(y+z)=xy+xz$	(Distributive)
11.	$x(y-z)=xy-xz$	(Distributive)
12.	$x^2=x$	(Idempotent)

These operations are similar to the modern laws of set theory with the set union operation represented by '+', and the set intersection operation is represented by multiplication. The universal set is represented by '1' and the empty by '0'. The associative and distributive laws hold. Finally, the set complement operation is given by $(1-x)$.

He applied the symbols to encode Aristotle's syllogistic logic, and he showed how the syllogisms could be reduced to equations. This allowed conclusions to be derived from premises by eliminating the middle term in the syllogism. He refined his ideas on logic further in *An Investigation of the Laws of Thought* which was published in 1854 [Boo:58]. This book aimed to identify the fundamental laws underlying reasoning in the human mind and to give expression to these laws in the symbolic language of a calculus.

He considered the equation $x^2=x$ to be a fundamental law of thought. It allows the principle of contradiction to be expressed (i.e. for an entity to possess an attribute and at the same time not to possess it):

$$\begin{aligned} x^2 &= x \\ \Rightarrow x - x^2 &= 0 \\ \Rightarrow x(1-x) &= 0 \end{aligned}$$

For example, if x represents the class of horses, then $(1-x)$ represents the class of 'not horses'. The product of two classes represents a class whose members are common to both classes. Hence, $x\ (1-x)$ represents the class whose members are at once both horses and 'not horses', and the equation $x\ (1-x)=0$ expresses the fact that there is no such class. That is, it is the empty set.

Boole contributed to other areas in mathematics including differential equations and finite differences[7] and to the development of probability theory. Des McHale has written an interesting biography of Boole [McH:85]. Boole's logic appeared to have no practical use, but this changed with Claude Shannon's 1937 master's thesis, which showed its applicability to switching theory and to the design of digital circuits.

[7] Finite differences are a numerical method used in solving differential equations.

3.6.1 Switching Circuits and Boolean Algebra

Claude Shannon showed in his famous master's thesis that Boolean algebra provided the perfect mathematical model for switching theory and for the design of digital circuits. It may be employed to optimize the design of systems of electromechanical relays, and circuits with relays solve Boolean algebra problems. The use of the properties of electrical switches to process logic is the basic concept that underlies all modern electronic digital computers. Digital computers today use the binary digits 0 and 1, and Boolean logical operations may be implemented by electronic AND, OR and NOT gates. More complex circuits (e.g. arithmetic) may be designed from these fundamental building blocks.

Modern electronic computers use billions of transistors that act as switches and can change state rapidly. The use of switches to represent binary values is the foundation of modern computing. A high voltage represents the binary value 1 with low voltage representing the binary value 0. A silicon chip may contain billions of tiny electronic switches arranged into logical gates. The basic logic gates are AND, OR and NOT. These gates may be combined in various ways to allow the computer to perform more complex tasks such as binary arithmetic. Each gate has binary value inputs and outputs.

The example in Fig. 3.7 is that of an 'AND' gate which produces the binary value 1 as output only if both inputs are 1. Otherwise, the result will be the binary value 0. Figure 3.8 shows an 'OR' gate which produces the binary value 1 as output if any of its inputs is 1. Otherwise, it will produce the binary value 0.

Finally, a NOT gate (Fig. 3.9) accepts only a single input which it reverses. That is, if the input is '1', the value '0' is produced and vice versa.

The logic gates may be combined to form more complex circuits. The example in Fig. 3.10 is that of a half-adder of 1 + 0. The inputs to the top OR gate are 1 and 0 which yields the result of 1. The inputs to the bottom AND gate are 1 and 0 which yields the result 0, which is then inverted through the NOT gate to yield binary 1. Finally, the last AND gate receives two 1s as input and the binary value 1 is the result of the addition. The half-adder (Fig. 3.10) computes the addition of two arbitrary binary digits, but it does not calculate the carry. It may be extended to a full adder that provides a carry for addition.

3.7 Application of Symbolic Logic to Digital Computing

Claude Shannon (Fig. 3.11) was an American mathematician and engineer who made fundamental contributions to computing. He was the first person[8] to see the applicability of Boolean algebra to simplify the design of circuits and telephone routing switches. He showed that Boole's symbolic logic developed in the nine-

[8] Victor Shestakov at Moscow State University also proposed a theory of electric switches based on Boolean algebra around the same time as Shannon. However, his results were published in Russian in 1941, whereas Shannon's were published in 1937.

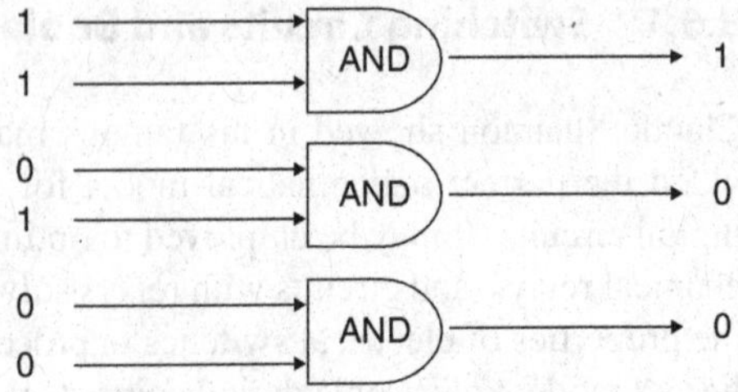

Fig. 3.7 Binary AND operation

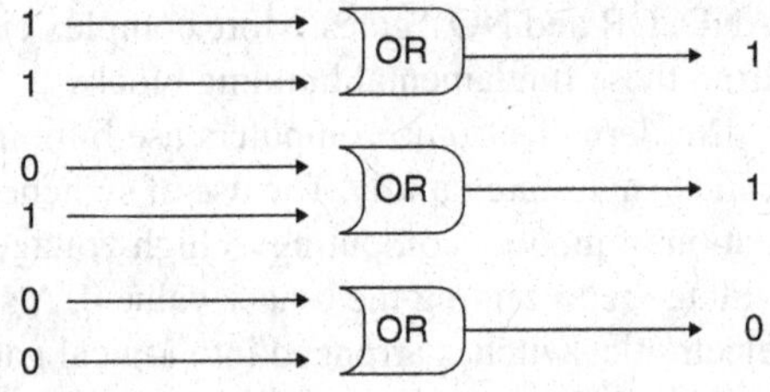

Fig. 3.8 Binary OR operation

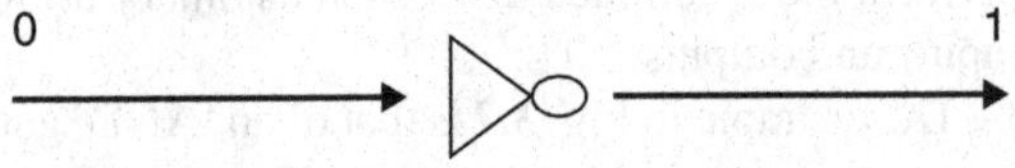

Fig. 3.9 NOT operation

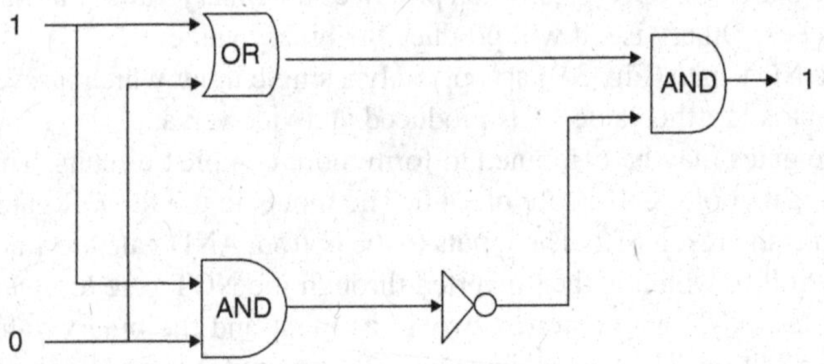

Fig. 3.10 Half-adder

teenth century provided the perfect mathematical model for switching theory and for the subsequent design of digital circuits and computers.

His influential *master's thesis is a key milestone in computing*, and it shows how to lay out circuits according to Boolean principles. It provides the theoretical foundation of switching circuits, and *his insight of using the properties of electrical switches to do Boolean logic is the basic concept that underlies all electronic digital computers*.

Shannon realized that you could combine switches in circuits in such a manner as to carry out symbolic logic operations. This allowed binary arithmetic and more complex mathematical operations to be performed by relay circuits. He designed a circuit, which could add binary numbers, and he later designed circuits that could make comparisons and thus are capable of performing a conditional statement. *This was the birth of digital logic and the digital computing age.*

Fig. 3.11 Claude Shannon

Vannevar Bush [ORg:13] was Shannon's supervisor at MIT, and Shannon's initial work was to improve Bush's mechanical computing device known as the differential analyser. This machine had a complicated control circuit that was composed of 100 switches that could be automatically opened and closed by an electromagnet. Shannon's insight was his realization that an electronic circuit is similar to Boolean algebra, and he showed how Boolean algebra could be employed to optimize the design of systems of electromechanical relays used in the analog computer. He also realized that circuits with relays could solve Boolean algebra problems.

He showed in his master's thesis *A Symbolic Analysis of Relay and Switching Circuits* [Sha:37] that the binary digits (i.e. 0 and 1) can be represented by electrical switches. The implications of true and false being denoted by the binary digits 1 and 0 were enormous, since it allowed binary arithmetic and more complex mathematical operations to be performed by relay circuits. This provided electronic engineers with the mathematical tool they needed to design digital electronic circuits and provided the foundation of digital electronic design.

The design of circuits and telephone routing switches could be simplified with Boolean algebra. Shannon showed how to lay out circuitry according to Boolean principles, and his master's thesis became the foundation for the practical design of digital circuits. These circuits are fundamental to the operation of modern computers and telecommunication systems, and his insight of using the properties of electrical switches to do Boolean logic is the basic concept that underlies all electronic digital computers.

He moved to the Mathematics Department at Bell Labs in the 1940s and commenced work that would lead to the foundation of modern *information theory*. The fundamental problem in this field is to reproduce at a destination point, either exactly or approximately, the message that has been sent from a source point. The problem is that information may be distorted by noise, leading to differences between the received message and the message that was originally sent. He provided a mathematical definition and framework for information theory in *A Mathematical Theory of Communication* [Sha:48]. He also contributed to the field of cryptography in *Communication Theory of Secrecy Systems* [Sha:49].

3.8 Review Questions

1. Explain the significance of binary numbers in the computing field.
2. Explain the importance of Shannon's master's thesis.
3. Explain the significance of the analytic engine.
4. Explain why Ada Lovelace is considered the world's first programmer.
5. Explain the significance of Boole to the computing field.
6. Explain the significance of Babbage to the computing field.
7. Explain the significance of Leibniz to the computing field.

3.9 Summary

This chapter considered foundational work done by Leibniz, Babbage, Boole, Ada Lovelace and Shannon. Leibniz was a seventeenth-century German mathematician and inventor who developed a calculating machine (the step reckoner) that could perform the four basic arithmetic operations. He also invented the binary number system, which is used extensively in the computer field.

Babbage was a nineteenth-century scientist and inventor who did pioneering work on calculating machines. He designed the difference engine (a sophisticated calculator that could be used for the production of mathematical tables), and he also designed the analytic engine (the world's first mechanical computer).

Lady Ada Lovelace was introduced into Babbage's ideas on the analytic engine, and she predicted that such a machine could be used to compose music, produce graphics as well as solve mathematical and scientific problems. She wrote what is considered the first computer program.

Boole was a nineteenth-century English mathematician who made important contributions to mathematics, probability theory and logic. Boole's symbolic logic provides the foundation for digital computers.

Shannon was a twentieth-century American mathematician and engineer, and he was the first person to see the applicability of Boolean algebra to simplify the design of circuits and telephone routing switches. He showed that Boole's symbolic logic provided the perfect mathematical model for switching theory and for the subsequent design of digital circuits and computers.

4 The First Digital Computers

Abstract
The Second World War motivated researchers to investigate faster ways to perform calculation to solve practical problems. This led to research into the development of digital computers to determine if they could provide faster methods of computation. We discuss the first digital computers including the Atanasoff-Berry computer developed in the United States, the ENIAC and EDVAC developed in the United States, the Colossus computer developed in England, Zuse's computers developed in Germany and the Manchester Mark I computer developed in England.

Key Topics
Harvard Mark I
ABC
ENIAC
EDVAC
Colossus
Zuse's machines
Manchester Mark I

4.1 Introduction

This chapter considers some of the early computers developed in the United States, Britain and Germany. The Second World War motivated researchers to investigate faster ways to perform calculation to solve practical problems. This led to research into the development of digital computers to determine if they could provide faster methods of computation.

G. O'Regan, *Introduction to the History of Computing*, Undergraduate Topics in Computer Science, DOI 10.1007/978-3-319-33138-6_4

The early computers were mainly large bulky machines consisting of several thousand vacuum tubes. A computer often took up the space of a large room, and it was slow and unreliable.

The early computers considered in this chapter include the Harvard Mark I designed and developed by Howard Aiken and IBM. This was a large electromechanical calculator that could perform mathematical calculations quickly. John Atanasoff and Clifford Berry designed and developed the Atanasoff-Berry computer (ABC), and this machine was designed to solve a set of linear equations using Gaussian elimination. John Mauchly and Presper Eckert designed the ENIAC and EDVAC. ENIAC was a fixed-program computer that needed to be physically rewired to solve different problems, but the EDVAC computer implemented the concept of a stored program. This meant that the program instructions could be stored in memory and that all that was required to carry out a new task was to load a new program into memory.

The team at Bletchley Park in England designed and developed the Colossus computer as part of their codebreaking work during the Second World War. This allowed them to crack the German Lorenz codes and to provide important military intelligence for the D-Day landings of 1944.

Konrad Zuse designed and developed the Z1, Z2 and Z3 machines in Germany. The Z3 was operational in 1941, and it was the world's first programmable computer.

4.2 Harvard Mark I

Howard Aiken (Fig. 4.1) made several important contributions to the early computing field. He showed that a large calculating machine could be built that would provide speedy solutions to mathematical problems.

His idea was to construct an electromechanical machine that could perform mathematical operations quickly and efficiently, and the machine would need to be able to handle positive and negative numbers and scientific functions, such as logarithms, and be able to work with minimal human intervention.

He discussed the idea with colleagues and IBM, and he was successful in obtaining IBM funding to build the machine. The machine was built at the IBM laboratories at Endicott with several IBM engineers involved in its construction. The construction took 7 years, and it was completed in 1943.

The machine became known as the Harvard Mark I (also known as the IBM *Automatic Sequence Controlled Calculator* (ASCC)). Aiken was influenced by Babbage's ideas on the design of the Analytic Engine.

IBM presented the machine to Harvard University in 1944, and the ASCC was essentially an electromechanical calculator that could perform large computations automatically. It could perform addition, subtraction, multiplication and division, and it could refer to previous results.

The Harvard Mark I (Fig. 4.2) was designed to assist in the numerical computation of differential equations, and it was 50 ft long and 8 ft high and weighed 5 tons.

Fig. 4.1 Howard Aiken

Fig. 4.2 Harvard Mark I (IBM ASCC) (Courtesy of IBM Archives)

It performed additions in less than a second, multiplications in 6 s and division in about 12 s. It used electromechanical relays to perform the calculations, and it could execute long computations automatically.

It was constructed out of switches, relays, rotating shafts and clutches, and it used 500 miles of wiring and over 750,000 components. It was the industry's largest electromechanical calculator, and it had 60 sets of 24 switches for manual data entry. It could store 72 numbers, each 23 decimal digits long. The instructions were read on paper tape, and punched cards were used to input the data, and the results were either on punched cards or an electric typewriter.

The US Navy used the Harvard Mark I for ballistic calculations, and the machine remained in use until 1959. It cost approximately half a million dollars, but it was never mass produced by IBM. It differed from most of the early digital computers in that it used relays instead of vacuum tubes.

The announcement of the Harvard Mark I led to tension between Aiken and IBM, as Aiken announced himself as the sole inventor without acknowledging the important role played by IBM.

4.3 Atanasoff-Berry Computer

John Atanasoff (Fig. 4.3) was born in New York in 1903, and he studied electrical engineering at the University of Florida and did a master's in mathematics at Iowa State College. He earned a PhD in theoretical physics from the University of

Fig. 4.3 John Atanasoff with components of ABC

Fig. 4.4 Replica of ABC (Courtesy of Iowa State University)

Wisconsin in 1930 and became an assistant professor at Iowa State College, where he taught mathematics and physics.

He became interested in developing faster methods of computation while doing his PhD research, as he wished to ease the time-consuming burden of calculation. He did some work on an analog calculator in 1936, but he concluded that analog devices were too restrictive and that they would be unable to give him the desired accuracy. His goal was to mechanize calculation to enable accurate computation to be carried out faster.

The existing computing devices were mechanical, electromechanical or analog. Atanasoff developed the concept of digital machine in the late 1930s, and he believed that his proposed machine would perform faster computations and be more accurate than the existing analog machines. He published the design of a machine to solve linear equations using his own version of Gaussian elimination in the summer of 1939. He then used his research grant of $650 to build the Atanasoff-Berry computer (ABC), with the assistance of his graduate student, Clifford Berry, from 1939 to 1942.

The ABC (Fig. 4.4) was approximately the size of a large desk and had approximately 270 vacuum tubes. Two hundred and ten tubes controlled the arithmetic unit; 30 tubes controlled the card reader and card punch; and the remaining tubes helped maintain charges in the condensers. It employed rotating drum memory, with each of the two drum memory units able to hold thirty 50-bit numbers.

The ABC was a digital machine that was designed for a specific purpose (i.e. solving linear equations) rather than as a general-purpose computer. The working prototype was one of the earliest electronic digital computers.[1] However, the ABC was slow, and it required constant operator monitoring.

It used binary mathematics and Boolean logic to solve simultaneous linear equations. It employed over 270 vacuum tubes for digital computation, but it had no central processing unit (CPU), and it was not programmable.

It weighed over 300 kg and it used 1.6 km of wiring. It used 50-bit numbers, and it could perform 30 additions or subtractions per second. The memory and arithmetic units could operate and store 60 such numbers at a time (60 * 50 = 3000 bits). The arithmetic logic unit was fully electronic, and it was implemented with vacuum tubes.

The input was in decimal format with standard IBM 80 column punched cards, and the output was in decimal format via a front panel display. A paper card reader was used as an intermediate storage device to store the results of operations too large to be handled entirely within electronic memory. The ABC pioneered important elements in modern computing including:

- Binary arithmetic and Boolean logic.
- All calculations were performed using electronics rather than mechanical switches.
- Computation and memory were separated.

The ABC was tested and operational by 1942, and its historical significance is that it demonstrated the feasibility of electronic computing. Several of its concepts were later used in the ENIAC developed by Mauchly and Eckert.

4.4 ENIAC and EDVAC

The Electronic Numerical Integrator and Computer (ENIAC) was one of the first large general-purpose digital computers. It was used to integrate ballistic equations and to calculate the trajectories of naval shells. It was completed in 1946, and it remained in use until 1955. The original cost of the machine was approximately $500,000.

ENIAC (Fig. 4.5) was a large bulky machine and it was over 100 ft long, 10 ft high and 3 ft deep and weighed about 30 tons. Its development commenced in 1943 at the University of Pennsylvania, and it was built for the US Army's Ballistics Research Laboratory The project team included Presper Eckert as chief engineer, John Mauchly as a consultant and several others. ENIAC had over 18,000 vacuum tubes, and so the machine generated a vast quantity of heat, as each vacuum tube

[1] The ABC was ruled to be the first electronic digital computer in the Sperry Rand vs. Honeywell patent case in 1973. However, Zuse's Z3 computer which was completed in Germany in 1941 preceded it.

Fig. 4.5 Setting the switches on ENIAC's function tables (US Army photo)

generated heat like a light bulb. The machine used 150 kW of power and air conditioning was used to cool it.

It employed decimal numerals and it could add five thousand numbers and do over three hundred and fifty 10-digit multiplications or thirty-five 10-digit divisions in one second. It could be programmed to perform complex sequences of operations, and this included loops, branches and subroutines. However, the task of taking a problem and mapping it onto the machine was complex, and it usually took weeks to perform. The first step was to determine what the program was to do on paper; the second step was the process of manipulating the switches and cables to enter the program into ENIAC, and this usually took several days. The final step was verification and debugging, and this often involved single-step execution of the machine.

There were problems initially with the reliability of ENIAC, as several vacuum tubes burned out most days (Fig. 4.6). This meant that the machine was often nonfunctional, as high-reliability vacuum tubes only became available in the late 1940s. However, most of these problems with the tubes occurred during the warm-up and cool-down periods, and it was therefore decided not to turn the machine off. This led to improvements in its reliability to the acceptable level of one tube every 2 days. The longest continuous period of operation without a failure was 5 days.

The very first program run on ENIAC took just 20 s, and the answer was manually verified to be correct after 40 h of work with a mechanical calculator. One of the earliest problems solved was related to the feasibility of the hydrogen bomb, and this program involved the input of 500,000 punched cards, and it ran for 6 weeks and gave an affirmative reply.

Fig. 4.6 Replacing a valve on ENIAC (US Army photo)

ENIAC was a *fixed-program* computer, and the machine had to be physically rewired in order to perform different tasks. It was clear that there was a need for an architecture that would allow a machine to perform different tasks without physical rewiring each time. This led to the concept of the *stored program*, which was implemented on EDVAC (the successor to ENIAC).

The idea of a stored program is that the program is stored in memory, and whenever there is a need to change the task that is to be computed, then all that is required is to place a new program in the memory of the computer, rather than rewiring the machine. EDVAC implemented the concept of a stored program in 1949, just after its implementation on the Manchester Baby prototype machine in England. The concept of a stored program and Von Neumann architecture is detailed in Von Neumann's report on EDVAC [VN:45].

ENIAC was preceded in development by Zuse's Z3 machine in Germany, the Atanasoff-Berry computer (ABC) in the United States and the Colossus computer developed in the United Kingdom. ENIAC was a major milestone in the history of computing.

4.4.1 EDVAC

The EDVAC (Electronic Discrete Variable Automatic Computer) was the successor to the ENIAC. It was a stored-program computer and it cost $500,000. Eckert and

Fig. 4.7 The EDVAC computer (US Army Photo)

Mauchly proposed it in 1944, and design work commenced prior to the completion of ENIAC.

It was delivered to the Ballistics Research Laboratory in 1949, and it commenced operations in 1951. It remained in operations until 1961. It employed 6000 vacuum tubes and its power consumption was 56,000 W. It had 5.5 Kb of memory.

EDVAC (Fig. 4.7) was one of the earliest stored-program computers, and the program instructions were stored in memory, rather than rewiring the machine each time.

4.4.2 Controversy Between the ABC and ENIAC

The ABC was ruled to be the first electronic digital computer in the 1963 *Honeywell vs. Sperry Rand* patent court case in the United States. The court case arose from a patent dispute between Sperry and Honeywell, and John Atanasoff was called as an expert witness in the case.

The court ruled that Eckert and Mauchly did not invent the first electronic computer, since the ABC existed as *prior art* at the time of their patent application. It is fundamental in patent law that an invention is novel and that there is no existing prior art. This meant that the Mauchly and Eckert patent application for ENIAC was invalid, and John Atanasoff was named as the inventor of the first digital computer.

Mauchly had visited Atanasoff on several occasions prior to the development of ENIAC, and they had discussed the implementation of the ABC. Mauchly subsequently designed the ENIAC, EDVAC and UNIVAC.

The court ruled that the ABC was the first digital computer and that the inventors of ENIAC had derived the subject matter of the electronic digital computer from Atanasoff.

4.5 Bletchley Park and Colossus

Tommy Flowers (Fig. 4.8) was a British engineer who made important contributions to breaking the Lorenz codes during the Second World War. He led the team that designed and built Colossus, which was one of the earliest electronic computers. The machine was designed to decode the top-level encrypted German military

Fig. 4.8 Tommy Flowers

communication sent by German High Command to its commanders in the field. This provided British and American intelligence with important information on German military plans around the D-Day invasion and later battles, and it helped to ensure the success of the Normandy landings and the ultimate defeat of Nazi Germany.

Flowers was born in East London in 1905, and he obtained a position with the telecommunications branch of the General Post Office in 1926. He moved to the research station at Dollis Hill in 1930, and he investigated the use of electronics for telephone exchanges. He was convinced at an early stage that an all-electronic system was possible.

He became involved with the codebreaking work taking place at Bletchley Park (located near Milton Keynes north west of London) during the Second World War. Alan Turing and others had cracked the German Enigma codes by building a machine known as the Bombe. This machine employed a crib to deduce the settings of the Enigma machine for that day. Turing introduced Flowers to Max Newman who was leading British efforts to break a German cipher generated by the Lorenz SZ42 machine.

Their existing approach to deciphering the Lorenz codes was with the Heath Robinson machine (a slow and unreliable machine). Flowers proposed an alternate solution involving the use of an electronic machine in 1943. This machine was called Colossus and it employed 1800 thermionic valves. The management at Bletchley Park were sceptical, but they encouraged him to continue with his work.

Flowers and others at the Post Office Research Station built the machine in 11 months, and its successor, the Colossus Mark 2, contained 2400 valves and it commenced operations on June 1, 1944. It was a large bulky machine and took up the space of a small room and weighed a ton.

It provided vital information for the Normandy landings, and it confirmed that Hitler had been successfully misled by Allied disinformation into believing that the Normandy landings were to be a diversionary tactic. Further, it confirmed that no additional German troops were to be moved there. The Colossus Mark 2 machine played a key role in helping the British to monitor the German reaction to their deception tactics.

4.5.1 Colossus

Flowers and others designed and built the original Colossus machine at the Post Office Research Station at Dollis Hill in London. The machine was used to find possible key combinations for the Lorenz machines rather than decrypting an intercepted message in its entirety. The Lorenz machine was based on the *Vernam cipher*.

Colossus compared two data streams to identify possible key settings for the Lorenz machine. The first data stream was the encrypted message, and it was read at high speed from a paper tape. The second stream was generated internally and was an electronic simulation of the Lorenz machine at various trial settings. If the match count for a setting was above a certain threshold, it would be sent as output to an electric typewriter.

Fig. 4.9 Colossus Mark 2 (Photo courtesy of UK government)

The Lorenz codes were a more complex cipher than the Enigma codes, and they were used in the transmission of important messages between the German High Command in Berlin and the military commanders in the field. The Lorenz SZ 40/42 machine performed the encryption. The Bletchley Park codebreakers called the typewriter-coding machine *Tunny* and the coded messages *Fish*. The codebreaking work involved carrying out complex statistical analyses on the intercepted messages.

The Colossus Mark I machine was specifically designed for codebreaking rather than as a general-purpose computer. It was semi-programmable and helped in deciphering messages encrypted using the Lorenz machine. A prototype was available in 1943 and a working version was available in the early 1944 at Bletchley Park. The Colossus Mark 2 (Fig. 4.9) was introduced just prior to the Normandy landings in June 1944.

The Colossus Mark I used 15 kW of power and it could process 5000 characters of paper tape per second. It enabled a large amount of mathematical work to be done in hours rather than in weeks. There were 10 Colossus machines working at Bletchley Park by the end of the war. A replica of the Colossus was rebuilt by a team of volunteers led by Tony Sale from 1993 to 1996, and it is at Bletchley Park museum.

The contribution of Bletchley Park to the cracking of the German Enigma and Lorenz codes and to the development of computing remained clouded in secrecy until recent times. The museum at Bletchley Park provides insight to the important contribution made by this organization to codebreaking and to early computing during the Second World War.

4.6 Zuse's Machines

Konrad Zuse is considered *the father of the computer* in Germany, as he built the world's first programmable machine (the Z3) in 1941 (Fig. 4.10).

He was born in Berlin in 1910, and he studied civil engineering at the Technical University of Berlin. He commenced working for Henschel (an airline manufacturer) after his graduation in 1935. He resigned after 1 year with the intention of forming his own company to build automatic calculating machines.

His parents provided financial support, and he commenced work on what would become the Z1 machine in 1936. Zuse employed the binary system for the calculator and metallic shafts that could shift from position 0 to 1 and vice versa. The Z1 was operational by 1938.

He served in the German Army on the Eastern Front for 6 months in 1939 at the start of the Second World War. Henschel helped Zuse to obtain a deferment from the army, and they made the case that he was needed as an engineer and not as a soldier. Zuse recommenced working at Henschel in 1940, and he remained affiliated with Henschel for the duration of the war. He built the Z2 and Z3 machines during this period, and the Z3 was operational in 1941, and it was the world's first programmable computer.

He started his own company in 1941, and this was the first company founded with the sole purpose of developing computers. The Z4 was almost complete as the Red Army advanced on Berlin in 1945, and Zuse left Berlin for Bavaria with the Z4 prior to the Russian advance. His other machines were destroyed in the Allied bombing of Germany.

He designed the world's first high-level programming language between 1943 and 1945, and this language was called Plankalkül. He later restarted his company

Fig. 4.10 Konrad Zuse (Courtesy of Horst Zuse, Berlin)

(Zuse KG), and he completed the Z4 in 1950. This was the first commercial computer, as it was completed ahead of the Ferranti Mark I, UNIVAC and LEO computers. Its first customer was the Technical University of Zurich.

Zuse's results are all the more impressive given that he was working alone in Germany, and he was unaware of the developments taking place in other countries. There is more detailed information on Zuse in [ORg:13].

4.6.1 Z1, Z2 and Z3 Machines

Zuse was unaware of computer-related developments in Germany or in other countries, and he independently implemented the principles of modern digital computers in isolation.

He commenced work on his first machine called the Z1 in 1936, and the machine was operational by 1938. It was demonstrated to a small number of people who saw it rattle and compute the determinant of a three by three matrix. It was essentially a binary electrically driven mechanical calculator with limited programmability. It was capable of executing instructions read from the program punched cards, but the program itself was never loaded into the memory.

It employed the binary system and metallic shafts that could slide from position 0 to position 1 and vice versa. The machine was essentially a 22-bit floating-point value adder and subtracter. A decimal keyboard was used for input, and the output was decimal digits. The machine included some control logic, which allowed it to perform more complex operations such as multiplications and division. These operations were performed by repeated additions for multiplication and repeated subtractions for division. The multiplication took approximately 5 s. The computer memory contained 64 22-bit words. Each word of memory could be read from and written to by the program punched cards and the control unit. It had a clock speed of 1 Hz and two floating-point registers of 22 bits each. The machine was unreliable, and a reconstruction of it is in the Deutsches Technikmuseum in Berlin.

His Z2 machine aimed to improve the Z1, and this mechanical and relay computer was created in 1939. It used a similar mechanical memory, but it replaced the arithmetic and control logic with 600 electrical relay circuits. It used 16-bit fixed-point arithmetic instead of the 22-bit used in the Z1. It had a 16-bit word size and the size of its memory was 64 words. It had a clock speed of 3 Hz.

The Z3 machine (Fig. 4.11) was the first functional tape-stored-program-controlled computer, and it was created in 1941. It used 2600 telephone relays and the binary number system, and it could perform floating-point arithmetic. It had a clock speed of 5Hz, and multiplication and division took 3 s. The input to the machine was with a decimal keyboard, and the output was on lamps that could display decimal numbers. The word length was 22 bits, and the size of the memory was 64 words.

It used a punched film for storing the sequence of program instructions. It could convert decimal to binary and back again. It was the first digital computer since it predates the Atanasoff-Berry computer by 1 year. It was proven to be Turing

Fig. 4.11 Zuse and the reconstructed Z3 (Courtesy of Horst Zuse, Berlin)

complete in 1998. There is a reconstruction of the Z3 computer in the Deutsches Museum in Munich.

4.7 University of Manchester

The Manchester Small-Scale Experimental Computer (better known by its nickname 'Baby') was developed at the University of Manchester. It was the *first stored-program computer*, and it was designed and built at Manchester University in England by Frederic Williams, Tom Kilburn, Geoff Tootill and others.

The machine demonstrated the reliability of the Williams tube, and it was the first stored-program computer: in other words the task to be computed is defined by the computer instructions that are placed in memory, and in order to change the task to be computed, all that is required is to load a different program into the computer memory. Kilburn wrote and executed the first stored program, and it was a short 17-line program written and executed in 1948.

The prototype 'Baby' (Fig. 4.12) demonstrated the feasibility and potential of a stored-program computer. Its memory consisted of 32 32-bit words, and it took 1.2 ms to execute one instruction: i.e. 0.00083 MIPS (million instructions per second). Today's computers are rated at speeds of up to 1000 MIPS and more. The team in Manchester developed the machine further, and in 1949, the Manchester Mark I was available.

Fig. 4.12 Replica of the Manchester Baby (Courtesy of Tommy Thomas)

4.7.1 Manchester Mark I

The Manchester Automatic Digital Computer (MADC), also known as the Manchester Mark I, was developed at the University of Manchester. It was one of the earliest stored-program computers, and it was the successor to the Manchester 'Baby' computer. It was designed and built by Williams, Kilburn and others.

Each word could hold one 40-bit number or two 20-bit instructions. The main memory consisted of two pages (i.e. two Williams tubes with each holding 32 × 40-bit words or 1280 bits). The secondary backup storage was a magnetic drum consisting of 32 pages (this was updated to 128 pages in the final specification). Each track consisted of two pages (2560 bits). One revolution of the drum took 30 ms, and this allowed the 2560 bits to be transferred to main memory.

The Manchester Mark I (Fig. 4.13) contained 4050 vacuum tubes, and it had a power consumption of 25,000 W. The standard instruction cycle was 1.8 ms but multiplication was much slower. The machine had 26 defined instructions, and the programs were entered into the machine in binary format, as assembly languages and assemblers were not yet available.

Fig. 4.13 The Manchester Mark I (Courtesy of the University of Manchester)

It had no operating system and its only systems software were some basic routines for input and output. Its peripheral devices included a teleprinter and a 5-hole paper tape reader and punch.

A display terminal used with the Manchester Mark I computer mirrored what was happening within the Williams tube. A metal detector plate placed close to the surface of the tube detected changes in electrical charges. The metal plate obscured a clear view of the tube, but the technicians could monitor the tubes used with a video screen. Each dot on the screen represented a dot on the tube's surface, and the dots on the tube's surface worked as capacitors that were either charged and bright or uncharged and dark. The information translated into binary code (0 for dark, 1 for bright) became a way to program the computer.

The Manchester Mark I influenced later computer development such as Ferranti Mark I general-purpose computer which was released in 1951, as well as early IBM computers such as the IBM 701.

4.8 Review Questions

1. Explain the significance of the ABC computer.
2. Explain what is meant by a 'stored-program' computer and its advantages over a fixed-program machine such as ENIAC.
3. Explain why Konrad Zuse is considered the father of the computer in Germany.
4. Explain the significance of the Manchester Baby computer.
5. Explain the significance of the work done at Bletchley Park during the Second World War.
6. Explain the significance of the Harvard Mark I?

4.9 Summary

This chapter considered some of the earliest computers developed in the United States, Britain and Germany. The Second World War led to research into the development of digital computers to determine if they could provide faster methods of computation. The early computers were mainly large bulky machines consisting of several thousand vacuum tubes. A computer often took up the space of a large room, and it was slow and unreliable.

The early computers considered in this chapter include the Harvard Mark I designed and developed by Howard Aiken and IBM. This was a large electromechanical calculator that could perform mathematical calculations quickly. Atanasoff and Berry designed and developed the ABC, and this machine was designed to solve a set of linear equations. Mauchly and Eckert designed the ENIAC and EDVAC, and the ENIAC needed to be physically rewired to solve different problems. Its successor, the EDVAC, implemented the concept of a stored program, which meant that a new program was loaded into the memory of the machine to solve a different problem.

The team at Bletchley Park in England designed and developed the Colossus computer as part of their codebreaking work. This allowed them to crack the German Lorenz codes and to provide important military information during the D-Day landings of 1944.

Konrad Zuse designed and developed the Z1, Z2 and Z3 machines in Germany. The Z3 was operational in 1941 and it was the world's first programmable computer. Williams, Kilburn and others implemented the first stored-program computer. This machine was popularly known as the Manchester Baby.

The First Commercial Computers

5

Abstract

This chapter discusses the first commercial computers including the UNIVAC I developed by EMCC/Sperry in the United States for the US Census Bureau and the LEO I computer developed by J. Lyons and Co. in England in partnership with Maurice Wilkes of Cambridge University. The Z4 computer was developed by Zuse KG in Germany, and the Ferranti Mark I computer was developed by Ferranti in partnership with the University of Manchester.

Key Topics

UNIVAC I

LEO I computer

Ferranti Mark I

Z4

CSIRAC

5.1 Introduction

This chapter considers a selection of the first commercial computers designed and developed in the United States, Great Britain, Germany and Australia. These machines built on the work of the first computers developed during the Second World War.

These include the UNIVAC I computer developed by EMCC (later called Sperry and Unisys) in the United States, the LEO I computer developed by J. Lyons and Co. in England, the Z4 computer developed by Zuse KG in Germany, the Ferranti Mark I developed by Ferranti in England and CSIRAC developed by CSIR in Australia.

G. O'Regan, *Introduction to the History of Computing*, Undergraduate Topics in Computer Science, DOI 10.1007/978-3-319-33138-6_5

The UNIVAC I computer was designed by John Mauchly and Presper Eckert of EMCC for the US Census Bureau, and it was designed for business and administrative use.

The LEO I computer was developed by J. Lyons and Co. in partnership with Cambridge University in England. It was based on the EDSAC computer designed by Maurice Wilkes at Cambridge University, and the LEO I was designed for business use.

The Z4 was designed and developed by Konrad Zuse in Germany. Zuse had already designed and developed a number of machines, and the Z4 computer was almost complete at the end of the Second World War. Zuse formed Zuse KG to complete the machine after the war.

The University of Manchester implemented the first stored-program computer (discussed in previous chapter), and the British government encouraged Ferranti to commercialize the Manchester Mark I.

5.2 UNIVAC

The Eckert-Mauchly Computer Corporation (EMCC) was founded by Presper Eckert and John Mauchly in 1947. It was one of the earliest computer companies in the world, and it pioneered a number of fundamental computer concepts such as the *stored program*, *subroutines*, *programming languages* and *compilers*.

EMCC was awarded a contract from the US Census Bureau in 1948 to develop the *Universal Automatic Computer* (UNIVAC) for the 1950 census. This was one of the first commercially available computers when it was delivered in 1951 (too late for the 1950 census), and it was designed for business and administrative use, rather than for complex scientific calculations. The UNIVAC machine was later used to accurately predict the result of the 1952 presidential election in the United States from a sample of 1 % of the population.

The UNIVAC I (Fig. 5.1) was initially priced at $159,000 and the price gradually increased over the years to reach between $1.2 and $1.5 million. Over 46 of these computers were built and delivered.

It employed magnetic tape for high-speed storage and it used 5200 vacuum tubes. It consumed 125 kW of electricity and it could carry out 1905 operations per second. It took up 400 square foot of space, and its main memory consisted of 1000 words of 12 characters. The input/output was via the operator's console, several tape drives and an electric typewriter.

UNIVAC is the name of a series of digital computers produced by EMCC and its successors (i.e. Remington Rand, Sperry and Unisys). The original model was the UNIVAC I (Universal Automatic Computer I). The successor models in the original UNIVAC series included the UNIVAC II, which was released in 1958, and the UNIVAC III, which was released by Sperry Rand in 1962.

EMCC set up a department to develop software applications for the UNIVAC computer, and it hired Grace Murray Hopper in 1949 as one of its first programmers. Hopper played an important role in the development of programming

Fig. 5.1 UNIVAC I computer

languages, and she made important contributions to the early development of compilers, programming language constructs, data processing and the COBOL programming language. For more information on Grace Murray Hopper, see [ORg:13].

EMCC was taken over by Remington Rand in 1950. Remington had a background in the production of typewriters, and *the Remington Typewriter was the first to use the* QWERTY *keyboard.* Remington's acquisition of EMCC allowed it to enter the electronics market, and EMCC became the UNIVAC division of Remington Rand. Sperry took over Remington Rand in 1955, and it became known as Sperry Rand (and later just Sperry).

5.3 LEO I Computer

J. Lyons and Co. was an innovative and forward-thinking company, and it was committed to finding ways to continuously improve to serve its customers better. It sent two of its executives to the United States shortly after the Second World War to evaluate new methods to improve business processes. These two executives came across the early computers that had been developed in the United States, including the ENIAC that had been developed by John Mauchly and others. They recognized the potential of these early machines for business data processing.

They also became aware during their visit to the United States that Maurice Wilkes and others at Cambridge University in England were working on the design of a computer based on the ideas detailed in Von Neumann's report. On their return to England, they visited Wilkes at Cambridge University, who was working on the design of the EDSAC computer. They were impressed by his ideas and technical knowledge and the potential of the planned EDSAC computer. They prepared a report for Lyon's board recommending that a computer designed for data processing

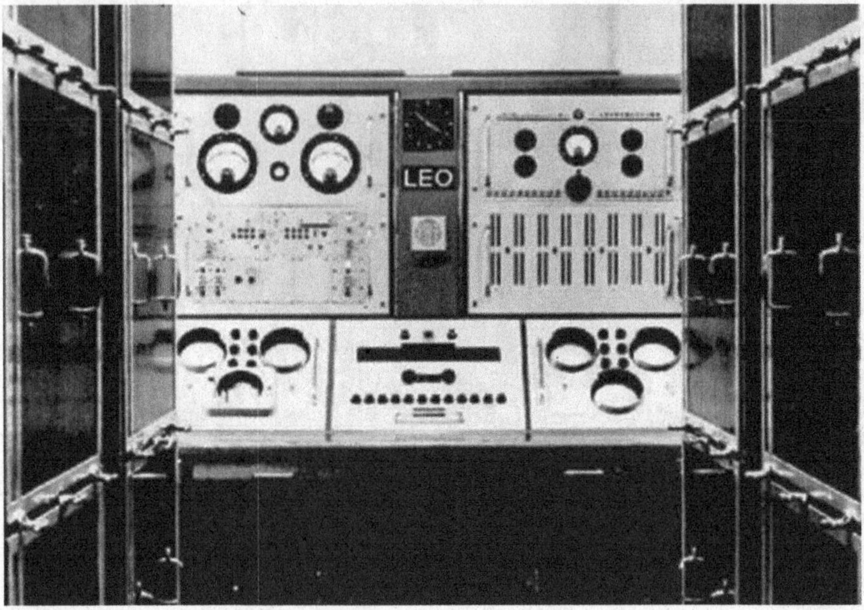

Fig. 5.2 LEO I computer (Courtesy of LEO Computer Society)

should be the next step in improving business processes and that Lyons should develop or acquire a computer to meet its business needs.

Lyons and Cambridge entered a collaboration arrangement where Lyons agreed to help fund the completion of EDSAC, and Cambridge agreed to help Lyons to develop its own computer, which was called the Lyons Electronic Office or LEO computer (Fig. 5.2). This machine was based on EDSAC but adapted to business data processing. Lyons set up a project team led by John Pinkerton to develop its computer, and Wilkes provided training for Lyon's engineers. The LEO computer ran its first program in late 1951.

The Electronic Delay Storage Automatic Calculator (EDSAC) was completed and ran its first program in 1949, and the LEO I computer was completed and ran its first program in late 1951. Lyons developed several applications for the LEO computer, and the LEO computer was used to process business applications (e.g. payroll) for other companies. Lyons recognized that more and more companies would require computing power, and they saw a business opportunity. They decided to set up a subsidiary company to focus on computers for commercial applications

Leo Computers Ltd. was set up in 1954 and it was based in London. It designed and developed a new computer, the LEO II, which was purchased by several British companies. The LEO III was released in 1961, and it was sold to customers in the United Kingdom and overseas.

LEO I's clock speed was 500 kHz with most instructions taking 1.5 ms to complete. The machine was linked to fast paper tape readers and fast punched card readers and punches. It had 8.75 Kb of memory holding 2048 35-bit words

The LEO I was initially used for valuation jobs, but this was later extended to payroll, inventory and other applications. One of the early applications developed by Lyons was an early version of an integrated management information system to manage its business. Lyons was also one of the pioneers of IT outsourcing in that it performed payroll calculations for a number of companies in the United Kingdom.

The UK Met Office used the LEO I computer in an early attempt at using a computer for weather forecasting in the early 1950s. The weather prediction model was solved on the LEO I computer, and the first predictions were made in 1954. The Met Office later used the Manchester Mark I and more powerful computers for weather forecasting. For a more detailed account of LEO, see [Fer:03, ORg:15].

5.4 The Z4 Computer

Zuse KG was founded by Konrad Zuse at Neukirchen (north of Frankfurt) in 1949. It was the first computer company in Germany, and it initially had five employees. The early focus of the company was to restore and improve Zuse's Z4 machine, which had survived the Allied bombing of Berlin and Zuse's subsequent move to Bavaria.

The Z4 machine (Fig. 5.3) consisted of 2200 relays, a mechanical memory of 64 32-bit words and a processor. The speed of the machine was approximate1000

Fig. 5.3 The Z4 computer (Courtesy of Horst Zuse, Berlin)

instructions per hour. The Henschel Aircraft Company had ordered the Z4 machine in 1942, but as the production of the machine was time consuming, it was never actually delivered to Henschel. The machine was almost complete by the end of the Second World War in 1945.

The Z4 was restored for the Institute of Applied Mathematics at the Eidgenössische Technische Hochschule (ETH) Zürich in Zurich. The restoration was complete in 1950, and it was delivered to the ETH later that year. It was one of the first operational computers in Europe at that time.

It was transferred to the French-German Research Institute of Saint-Louis in France in 1955, and it remained operational there until 1959. Today, the Z4 machine is on display at the Deutsche Museum in Munich.

Zuse KG commenced work on the Z5 in the early 1950s, and this was an extended version of the Z4. The Z5 was one of the first commercial computers in Europe, and it was produced for the Leitz company in Germany. The Z5 followed similar construction principles as the Z4, but it was over six times faster.

Zuse KG produced over 250 computers from 1949 to 1969, and by 1964 it had over 1200 employees. The company ran into financial difficulties in the early 1960s, and it was taken over by Rheinstahl in 1964. Rheinstahl was taken over by Siemens in 1967, and Konrad Zuse left the company in 1969. For a more detailed account of Zuse, see [ORg:15].

5.5 Ferranti Mark I

Ferranti Ltd. (a British company) and Manchester University collaborated to build one of the earliest general-purpose electronic computers. The machine was called the Ferranti Mark I (it was also known as the Manchester Electronic Computer), and it was basically an improved version of the Manchester Mark I.

The first machine off the production line was delivered to the University of Manchester in 1951 and shortly before the release of the UNIVAC I electronic computer in the United States.

The main improvements of the Ferranti Mark I over the Manchester Mark I computer were in the size of primary and secondary storage, a faster multiplier and additional instructions. The Ferranti Mark I (Fig. 5.4) had eight pages of random access memory (i.e. eight Williams tubes each with a storage capacity of 64 20-bit words or 1280 bits). A 512-page magnetic drum, which stored two pages per track, provided the secondary storage, and its revolution time was 30 ms.

It used a 20-bit word stored as a single line of dots on the Williams tube display, with each tube storing a total of 64 lines of dots (or 64 words). Instructions were stored in a single word, while numbers were stored in two words.

The accumulator was 80 bits and it could also be addressed as two 40-bit words. There were about 50 instructions and the standard instruction time was 1.2 ms. Multiplication could be completed in 2.16 ms. There were 4050 vacuum tubes employed.

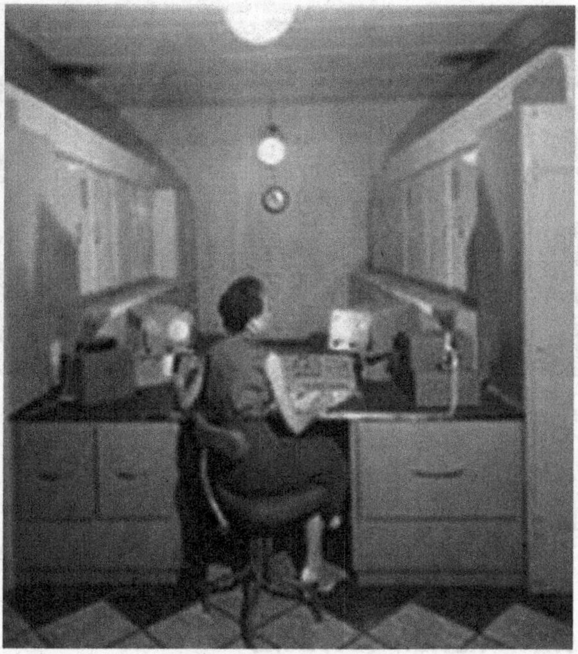

Fig. 5.4 Ferranti Mark I

The Ferranti Mark I's instruction set included a *hoot command* which allowed auditory sounds to be produced. It also allowed variations in pitch. Christopher Strachey (who later did important work in the semantics of programming languages) programmed the Ferranti Mark I to play tunes such as *God Save the King*, and the Ferranti Mark I was one of the earliest computers to play music.

Dr. Dietrich Prinz wrote one of the earliest computer games (a chess-playing program) for the Ferranti Mark I in 1951. The parents of Tim Berners-Lee (the inventor of the World Wide Web) both worked on the Ferranti Mark I.

5.6 CSIRAC Computer

The CSIRAC (Council for Scientific and Industrial Research Automatic Computer) was Australia's first digital computer. It was one of the earliest stored-program computers, and it became operational in November 1949. It is on permanent display at the Melbourne Museum.

It was constructed by a team led by Trevor Pearcey and Maston Beard at the CSIR in Sydney. The machine had 2000 vacuum valves and used 30 kW of power during operation. The input to the machine was done with a punched paper tape, and output was to a teleprinter or to punched tape. The machine was controlled through a console, which allowed programs to be stepped through one at a time.

Fig. 5.5 CSIRAC (Photo courtesy of John O'Neill)

The CSIRAC (Fig. 5.5) was the first digital computer to play music and this took place in 1950. The machine was moved to the University of Melbourne in the mid-1950s, and today the machine is on permanent display at the Melbourne Museum.

5.7 Review Questions

1. What are the key contributions made by EMCC/Unisys to the computing field ?
2. Describe the contributions of J. Lyons and Co. to the early computing field.
3. What is the significance of Zuse's Z4 machine?
4. Discuss the progress made in the production of music on early computers.
5. Describe the contribution of the University of Manchester to early computing. What were the key improvements in the Ferranti Mark I over the Manchester Mark I?
6. Describe the contributions of Grace Murray Hopper to the computing field.

5.8 Summary

This chapter considered a selection of the first commercial computers designed and developed in the United States, Britain, Germany and Australia. These machines built upon the work done on the first digital computers developed during the Second World War.

We discuss the UNIVAC I computer developed by EMCC in the United States, the LEO I computer developed by J. Lyons and Co. in England, the Z4 computer developed by Zuse KG in Germany, the Ferranti Mark I developed by Ferranti in England and CSIRAC developed by CSIR in Australia.

Mauchly and Eckert wished to commercialize their work on the ENIAC/EDVAC computers and to protect their intellectual property, as new policies at the University of Pennsylvania required them to sign over the intellectual property rights to their invention. They set up EMCC to commercialize their inventions.

The LEO I computer arose as a result of forward thinking by J. Lyons and Co. who wished to improve their businesses processes, and they collaborated with Maurice Wilkes at Cambridge University to produce the LEO I computer.

The UK government encouraged Ferranti to commercialize the Manchester Mark I computer, and the Ferranti Mark I improved upon it with larger primary and secondary storage and additional instructions.

Early Commercial Computers and the Invention of the Transistor

6

Abstract

This chapter considers a selection of computers developed during the 1950s, and it includes a selection of vacuum tube-based computers as well as transistor computers. One of the drivers for the design and development of more powerful computers was the perceived threat of the Soviet Union. This led to an arms race between the two superpowers, and it was clear that computing technology would play an important role in developing sophisticated weapon and defence systems. The SAGE air defence system developed for the United States and Canada was an early example of the use of computer technology for the military. Early IBM computers such as the IBM 701 and 704 computers are discussed, and the chapter concludes with a discussion of the invention of the transistor by William Shockley and others at Bell Labs.

Key Topics

IBM 701
SAGE
Transistor
IBM 608
IBM 704

6.1 Introduction

This chapter considers a selection of computers developed during the 1950s, and it includes a selection of vacuum tube-based computers as well as transistor computers. One of the drivers for the design and development of more powerful computers was the perceived threat of the Soviet Union. This led to an arms race between the

G. O'Regan, *Introduction to the History of Computing*, Undergraduate Topics in Computer Science, DOI 10.1007/978-3-319-33138-6_6

two superpowers, and it was clear that computing technology would play an important role in developing more sophisticated weapon and defence systems. The SAGE air defence system developed for the United States and Canada was an early example of the use of computer technology for the military.

The other key driver for the development of more powerful computers was to support business, universities and government. The machines developed during this period were mainly large proprietary mainframes designed for business, scientific and government use. They were expensive and this eventually led vendors such as IBM and DEC to introduce families of computers in the 1960s, where a customer could choose a small cheaper member of the family and to upgrade over time to a larger computer as their needs evolved.

The origins of IBM are in the work done by Herman Hollerith in developing a tabulating machine to process the 1890 census of the population of the United States. IBM became a very successful international company selling punched card tabulating machines. Thomas Watson Sr. led the company from 1912 to 1952, and Thomas Watson Jr. became CEO in 1952. He believed that the future of IBM was in computers, and not tabulators, and he transformed IBM to become a world leader in the computer sector.

6.2 Early IBM Computers

IBM commenced work on computers during the Second World War, with its joint venture with Howard Aiken on the Harvard Mark I (also known as the IBM Automatic Sequence Controlled Calculator (ASCC)). This machine was essentially an electromechanical calculator that could perform large computations automatically. We discussed this machine in Chap. 4, and it was delivered to Harvard University in 1941.

IBM introduced the Vacuum Tube Multiplier in 1943, which was an important move from electromechanical to electronic machines (the Harvard Mark I used electromechanical relays to perform the calculations). It was one of the first complete machines to perform arithmetic electronically by substituting vacuum tubes for electric relays. The key advantages of the vacuum tubes were that they were faster, smaller and easier to replace than the electromechanical switches used on the Harvard Mark I. This allowed engineers to process information thousands of times faster.

IBM introduced its first large computer based on vacuum tubes in 1952. The machine was called the IBM 701 (Fig. 6.1), and it executed 17,000 instructions per second. It was used mainly for government work and for business applications.

IBM introduced the IBM 650 (Magnetic Drum Calculator) in 1954. This was an intermediate-sized electronic computer designed to handle accounting and scientific computations. It was one of the first mass-produced computers, and universities and businesses used it. It was a very successful product for IBM, with over 2000 machines built and sold between its product launch in 1954, and its retirement in

Fig. 6.1 IBM 701 (Courtesy of IBM Archives)

1962. The machine included a central processing unit, a power unit and a card reader.

The IBM 704 data processing system (Fig. 6.2) was a large computer introduced in 1954. It included core memory and floating-point arithmetic, and it was used for scientific and commercial applications. It included high-speed memory which was faster and much more reliable than the cathode-ray-tube memory storage mechanism used in earlier machines. It also had a magnetic drum storage unit, which could store parts of the program and intermediate results.

The interaction with the system was either by magnetic tape or punched cards entered through the card reader. The program instructions or data were initially produced on punched cards. They were then either converted to magnetic tape or read directly into the system, and the data processing was then performed. The output from the data processing was then sent to a line printer, magnetic tape or punched cards. Multiplication and division was performed in 240 microseconds.

The designers of the IBM 704 included John Backus and Gene Amdahl. Backus was one of the key designers of the FORTRAN programming language, which was introduced by IBM in 1957. This was the first scientific programming language, and it is still popular with engineers and scientists. Gene Amdahl later founded Amdahl Corporation after his resignation from IBM, and Amdahl Corporation later became a major rival to IBM in the mainframe market. For more detailed information on Backus and Amdahl, see [ORg:13].

Fig. 6.2 IBM 704 (Courtesy of IBM Archives)

6.3 The SAGE System

The Semi-Automatic Ground Environment (SAGE) was an automated system for tracking and intercepting enemy aircraft in North America. It was used by the North American Aerospace Defense Command (NORAD), which is located in Colorado in the United States. The SAGE system was used from the late 1950s until the 1980s.

The interception of enemy aircraft was extremely difficult prior to the invention of radar during the Second World War. Its introduction allowed fighter aircraft to be scrambled just in time to meet the enemy threat. The radar stations were ground based, and they therefore needed to communicate with and send interception instructions to the fighter aircraft to deal with the hostile aircraft.

However, after the war the speed of aircraft increased considerably, thereby reducing the time available to scramble fighter aircraft. This necessitated a more efficient and automatic way to transmit interception instructions and new approaches to provide security for the United States. The SAGE system (Fig. 6.3) was designed to solve this problem, it analysed the information that it received from the various radar stations around the country in real time, *and it then automated the transmission of interception messages to fighter aircraft.*

IBM and MIT played an important role in the design and development of SAGE. Some initial work on real-time computer systems had been done at the Massachusetts Institute of Technology on a project for the US Navy. This project was concerned with building an aircraft flight simulator computer for training bombing crews, and it led to the development of the Whirlwind digital computer. This computer was originally intended to be an analog machine, but instead it became the Whirlwind digital computer, and it was used for experimental development of military combat information systems.

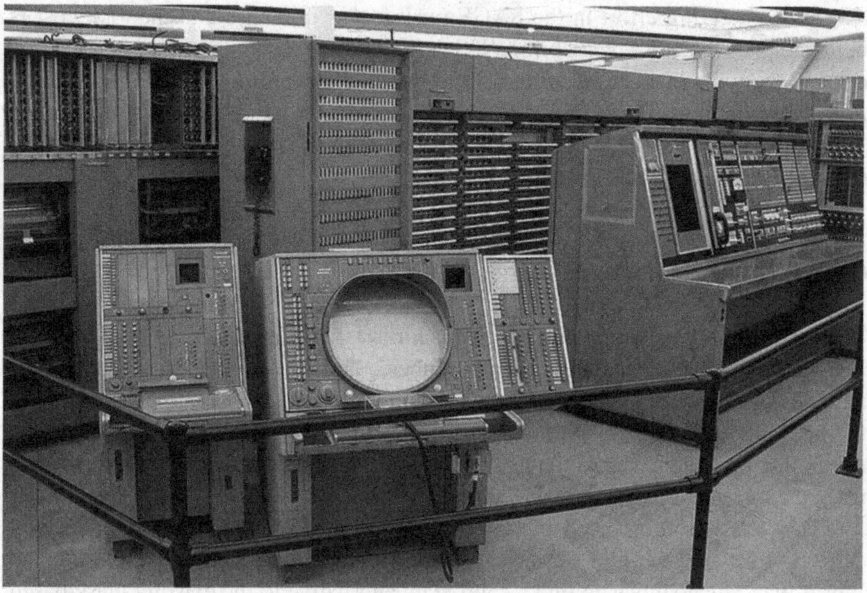

Fig. 6.3 SAGE (Photo courtesy of Steve Jurvetson)

Whirlwind was the first real-time computer, and George Valley and Jay Forrester wrote a proposal to employ Whirlwind for air defence. This led to the Cape Cod system, which demonstrated the feasibility of an air defence system covering New England. The design and development of SAGE commenced in 1953

IBM was responsible for the design and manufacture of the AN/FSQ-7 vacuum tube computer used in SAGE. Its design was based on the Whirlwind II computer, which was intended to be the successor to Whirlwind. However, the Whirlwind II was never built, and the AN/FSQ-7 computer weighed 275 tons and included 500,000 lines of assembly code.

The AN/FSQ holds the current world record for the largest computer ever built. It employed 55,000 vacuum tubes and covered an area over 18,000 square feet; and it used about 3 MW of power.

There were twenty-four SAGE Direction Centers and three SAGE Combat Centers located in the United Sates. Each SAGE site included two computers for redundancy, and long-distance telephone lines linked each centre. Burroughs provided the communications equipment to enable the centres to communicate with one another, and *this was one of the earliest computer networks*.

Each site was connected to multiple radar stations with tracking data transmitted by modem over a standard telephone wire. The SAGE computers then collected the tracking data for display on a cathode ray tube (CRT). The console operators at the centre could select any of the targets on the display to obtain information on the tracking data. This enabled aircraft to be tracked and identified, and the electronic information was presented to operators on a display device.

The engineering effort in the SAGE project was immense and the total cost is believed to have been several billion US dollars. It was a massive construction project, which involved erecting buildings and building power lines and communication links between the various centres and radar stations.

SAGE influenced the design and development of the Federal Aviation Authority (FAA) automated air traffic control system.

6.4 Invention of the Transistor

The early computers were large bulky machines taking up the size of a large room. They contained thousands of vacuum tubes,[1] and these tubes consumed large amounts of power and generated a vast quantity of heat. This led to problems with the reliability of the early computers, as several tubes burned out each day. This meant that machines were often non-functional for parts of the day, until the defective tube was identified and replaced (see Fig. 4.6).

There was therefore a need to find a better solution to vacuum tubes, and Shockley (Fig. 6.4) set up the solid physics research group at Bell Labs after the Second World War. His goal was to find a solid-state alternative to the existing glass-based vacuum tubes.

Fig. 6.4 William Shockley (Courtesy Chuck Painter, Stanford news service)

[1] ENIAC contained over 18,000 vacuum tubes and the AN/FSQ-7 computer used in SAGE contained 55,000 vacuum tubes.

Fig. 6.5 Replica of transistor (Courtesy of Lucent Bell Labs)

Shockley was born in England in 1910 to American parents, and he grew up at Palo Alto in California. He earned his PhD from the Massachusetts Institute of Technology in 1936, and he joined Bell Labs shortly afterwards. His solid physics research team included John Bardeen and Walter Brattain, who would later share the 1956 Nobel Prize in Physics with him for their invention of the transistor (Fig. 6.5).

Their early research was unsuccessful, but by late 1947 Bardeen and Brattain succeeded in creating a point-contact transistor independently of Shockley, who was working on a junction-based transistor. Shockley believed that the point-contact transistor would not be commercially viable, and his junction point transistor was announced in mid-1951, with a patent granted later that year. The junction point transistor soon eclipsed the point-contact transistor, and it became dominant in the marketplace.

Shockley published a book on semiconductors in 1950 [Sho:50], and he resigned from Bell Labs in 1955. He formed Shockley Laboratory for Semiconductors (part of Beckman Instruments) at Mountain View in California. This company played an important role in the development of transistors and semiconductors, and several of its staff later formed semiconductor companies in the Silicon Valley area.

Shockley was the director of the company, but his management style alienated several of his employees. This led to the resignation of eight key researchers in 1957 following his decision not to continue research into silicon-based semiconductors. This gang of eight went on to form Fairchild Semiconductors and other companies in the Silicon Valley area in the following years.

For more detailed information on Shockley and Bell Labs, see [ORg:13, ORg:15].

6.5 Early Transistor Computers

The University of Manchester Experimental Transistor Computer was one of the first transistor computers.[2] The prototype machine used 92 point-contact transistors and had a 48-bit word size, whereas the full-scale version used 200 point-contact transistors. There were serious problems with the reliability of the point-contact transistors, which meant that there were reliability problems with the machine. Metropolitan-Vickers (a Manchester company) adapted the design and changed the circuits to use the more reliable junction-based transistors and created a full-scale version called the Metrovick 950 in 1956.

Other early transistor computers include the TRADIC designed and developed by Bell Labs in the early 1954. This machine also used some vacuum tubes. The Harwell CADET was an early fully transistorized machine when it appeared in early 1955. The IBM 608 was the first IBM product to use transistor circuits instead of vacuum tubes. The prototype of this product appeared in 1955, and the fully transistorized calculator was introduced in late 1957. It contained 3000 germanium transistors. The Burroughs SM-65 Atlas ICBM was an early transistorized computer, which appeared in 1957.

The IBM 7090 was one of the earliest commercial computers with transistor logic, and it was introduced in 1958. It was designed for large-scale scientific applications, and it was over 13 times faster than the older vacuum tube IBM 701. It used 36-bit words, had an address space of 32,768 words and could perform 229,000 calculations per second. It was used by the US Air Force to provide an early warning system for missiles and also by NASA to control space flights. It cost approximately $3 million, but it could be rented for over $60 K per month.

6.6 Review Questions

1. Explain the significance of the transistor in the computing field.
2. Explain the significance of the SAGE system to the computing field.
3. Describe the contributions made by the University of Manchester to the computing field.
4. Describe the early transistor computers.
5. Describe the contributions of John Backus and Gene Amdahl to the computing field.
6. Describe the contributions of Bell Labs to the computing field.
7. Describe the contributions of IBM to the computing field.

[2] It was not a fully transistorized computer in that it employed a small number of vacuum tubes in its clock generator.

6.7 Summary

This chapter considers a selection of computers developed during the 1950s, and it includes a selection of vacuum tube-based computers as well as early transistor computers.

Among the early vacuum tube computers considered were the IBM 701 and IBM 704. The IBM 701 was introduced in 1952; it was used mainly for government work and for business applications. The IBM 704 data processing system was a large computer that was introduced in 1954. It was used for scientific and commercial applications, and Gene Amdahl and John Backus were involved in its design.

The SAGE air defence system was developed for the United States and Canada, and it was an early example of the use of computer technology for the military. It was an automated system for tracking and intercepting enemy aircraft in North America, and it automated the transmission of interception messages to fighter aircraft.

The invention of the transistor by Shockley and others at Bell Labs was a revolution in computing, and it led to smaller, faster and more reliable computers. The University of Manchester Experimental Transistor Computer was one of the earliest transistor computers.

The Invention of the Integrated Circuit and the Birth of Silicon Valley

7

Abstract

The invention of the integrated circuit allowed many transistors to be combined on a single chip, and it was another revolution in computing. The integrated circuit placed the previously separated transistors, resistors, capacitors and wiring circuitry onto a single chip made of silicon or germanium. The integrated circuit shrunk the size and cost of making electronics, and it had a major influence on the design of later computers leading to faster and more powerful machines. The germanium-based integrated circuit was invented by Jack Kilby at Texas Instruments, and Robert Noyce of Fairchild Semiconductor did subsequent work on silicon-based integrated circuits. Moore's law on the exponential growth of transistor density on an integrated circuit is discussed, as well as its relevance to the computing power of electronic devices.

Key Topics

Integrated circuit
Silicon
Germanium
Texas Instruments
Fairchild
Silicon Valley
Moore's law

G. O'Regan, *Introduction to the History of Computing*, Undergraduate Topics in Computer Science, DOI 10.1007/978-3-319-33138-6_7

7.1 Introduction

The first computers used thousands of vacuum tubes, and they were large bulky machines. The invention of the transistor was a revolution in computing, and it led to smaller, faster and more reliable computers. However, it was still a challenge for engineers to design complex circuits, as they had to wire hundreds (thousands) of separate components together.

It is essential when building a circuit that all of the connections are intact, as otherwise the electric current will be stopped on its way through the circuit, and the circuit will fail. Prior to the invention of the integrated circuit, engineers had to construct circuits by hand, which involved soldering each component in place and connecting them with wires. However, the manual assembly of the large number of components required in a computer often resulted in faulty connections, and advanced computers required so many connections that they were almost impossible to build. Clearly, there was a need for a better solution.

The invention of the integrated circuit allowed many transistors to be combined on a single chip, and it was another revolution in computing. The integrated circuit placed the previously separated transistors, resistors, capacitors and wiring circuitry onto a single chip made of silicon or germanium. The integrated circuit shrunk the size and cost of making electronics, and it had a major influence on the design of later computers and electronics. It led to faster and more powerful computers.

7.2 Invention of Integrated Circuit

The electronics industry was dominated by vacuum tube technology up to the mid-1950s. However, vacuum tubes had inherent limitations as they were bulky and unreliable, produced considerable heat and consumed a lot of power. Bell Labs invented the transistor in the late 1940s, and transistors were tiny in comparison to vacuum tubes and consumed very little power, and they were more reliable and lasted longer. The transistor stimulated engineers to design ever more complex electronic circuits and equipment containing hundreds or thousands of discrete components such as transistors, diodes, rectifiers and capacitors.

The motivation for the invention of the integrated circuit was the goal of finding a solution to the problems that engineers faced in increasing the performance of their designs as the number of components in the design increased. Each component needed to be wired to many other components, and the wiring and soldering was done manually. Clearly, more components would be required to improve performance, and therefore, it seemed that future designs would consist almost entirely of wiring.

However, the problem was that these components needed to be interconnected to form electronic circuits, and this involved hand soldering of thousands of components to thousands of bits of wire. This was expensive and time-consuming, and it was also unreliable since every soldered joint was a potential source of trouble. The

Fig. 7.1 Jack Kilby c. 1958 (Courtesy of Texas Instruments)

challenge for the industry was to find a cost-effective and reliable way of producing these components and interconnecting them.

Jack Kilby (Fig. 7.1) joined Texas Instruments in 1958, and he began investigating how to solve this problem. He realized that semiconductors were all that were really required, as resistors and capacitors could be made from the same material as the transistors. He realized that since all of the components could be made of a single material, they could also be made in situ interconnected to form a complete circuit.

Kilby succeeded in building an integrated circuit made of germanium that contained several transistors in 1958. Robert Noyce of Fairchild Semiconductors built an integrated circuit on a single wafer of silicon in 1960, and Kilby and Noyce are considered coinventors of the integrated circuit. Kilby was awarded the Nobel Prize in Physics in 2000 for his role in the invention of the integrated circuit.

Kilby's integrated circuit consisted of a transistor and other components on a slice of germanium (Fig. 7.2). His invention revolutionized the electronics industry, and the integrated circuit is the foundation of almost every electronic device in use today. His invention used germanium, and the size of the integrated circuit was 7/16 by 1/16 inches.

Robert Noyce at Fairchild Semiconductors later invented an integrated circuit based on a single wafer of silicon in 1960, and today silicon is the material of choice for semiconductors. Noyce made an important improvement on Kilby's design in that he added a thin layer of metal to the chip to better connect the various components in the circuit. Noyce's solution made the integrated circuit more suitable for mass production, and Fairchild Semiconductors pioneered the use of the *planar process* for making transistors, and the existing semiconductor companies soon

Fig. 7.2 First integrated circuit (Courtesy of Texas Instruments)

employed this process. Noyce was one of the co-founders of Intel, which is one of the largest manufacturers of integrated circuits in the world.

An integrated circuit (IC) consists of a set of electronic circuits on a small chip of semiconductor material, and it is much smaller than a circuit made out of independent components. The IC is made on a small plate of semiconductor material that is usually made of silicon. An integrated circuit is extremely compact, and it may contain billions of transistors and other electronic components in a tiny area. The width of each conducting line has got smaller and smaller due to advances in technology over the years, and it is now measured in tens of nanometres.[1] The invention of the integrated circuit led to major reductions in the size and cost of making electronics, and it impacted the design of all future computers and other electronics.

The size of the components in a modern fabrication plant is extremely small, with thousands of transistors fitting inside the cross section of a strand of hair. The production of a chip requires precision at the atomic level, with tiny particles such as those in tobacco smoke large enough to ruin a chip. For this reason, chip production takes place in a clean room, which is a special room designed with furniture made of special materials that don't give off particles and very effective air filters and air circulation systems.

There has been a massive reduction in the production costs of integrated circuits, with the initial production cost of integrated circuits at $1000 in 1960. However, as

[1] 1 nanometre (nm) is equal to 10^{-9} m.

demand increased and production techniques improved, the cost of production was reduced down to $25 by 1963.

There are several generations of integrated circuits from the small-scale integration (SSI) of the early 1960s, which typically had less than 30 transistors on the chip, to medium-scale integration (MSI) of the late 1960s with less than 300 transistors on the chip, to large-scale integration (LSI) of the mid-1970s with less than 3000 transistors on the chip, to very large-scale (VLSI) and ultra large-scale integration (ULSI) of the 1980s, which have over a million transistors on the chip.

There are several large companies that design and make semiconductors. These include companies such as Texas Instruments (TI), which is an American electronics company that is one of the largest manufacturers of semiconductors in the world. Intel and AMD (Advanced Micro Devices) are among the largest makers of semiconductors in the world. For more detailed information on Jack Kilby and Texas Instruments, see [ORg:13, ORg:15].

7.2.1 Moore's Law

Gordon Moore observed that over a period of time (from 1958 up to 1965), the number of transistors on an integrated circuit doubled approximately every year. This led him to formulate what became known as *Moore's law* in 1965 [Mor:65], which predicted that this trend would continue for at least another 10 years. He refined the law in 1975 and predicted that a doubling in transistor density would occur every 2 years for the following 10 years.

His prediction of *exponential growth* in transistor density has proved to be accurate over the last 50 years, and the capabilities of many digital electronic devices are linked to Moore's law.

The exponential growth in the capability of processor speed, memory capacity and so on is all related to this law. It is likely that the growth in transistor density will slow to a doubling of density every 3 years by 2015.

The phenomenal growth in productivity is due to continuous innovation and improvement in manufacturing processes. It has led to more and more powerful computers running more and more sophisticated applications.

7.3 Early Integrated Circuit Computers

It took some time for integrated circuits to take off, as they were an unproven technology and they remained expensive until mass production. Kilby and others at Texas Instruments successfully commercialized the integrated circuit by designing a hand-held calculator that was as powerful as the existing large, electromechanical desktop models. The resulting electronic hand-held calculator was small enough to fit in a coat pocket. This battery-powered device could perform the four basic arithmetic operations on six-digit numbers, and it was completed in 1967.

Fig. 7.3 The DEC PDP-8/e

The earliest computers that used integrated circuits appeared in the 1960s, and the early use of integrated circuits was mainly in embedded systems. The use of integrated circuits played an important role in early aerospace projects such as the Apollo Guidance Computer and Minuteman missile. The Apollo flight computer was one of the earliest computers to use integrated circuits, and it was developed by MIT/Raytheon and introduced in 1966. It provided capabilities for the guidance, navigation and control of the Apollo spacecraft. The Minuteman II program used a computer built from integrated circuits, and the guidance system of the Minuteman II intercontinental ballistic missile was much smaller due to the use of the integrated circuits.

DEC's first minicomputer to use integrated circuits was the popular PDP-8 (Fig. 7.3), which was designed by Edson de Castro and introduced in 1965. Hewlett-Packard introduced the 2116A minicomputer in 1966, and this minicomputer used Fairchild Semiconductors integrated circuits.

The Honeywell ALERT airborne computer was designed to handle complex airborne data in a real-time environment, and it was introduced in 1966. The Central Air Data Computer was designed in the late 1960s, and it was used for flight control in the US Navy's F-14A Tomcat fighter. These were among the early computers to use integrated circuits.

7.4 Birth of Silicon Valley

Silicon Valley is the nickname for the southern portion of the San Francisco Bay Area. It is home to many of the world's largest high-tech companies as well as thousands of start-up companies.

Fig. 7.4 HP Palo Alto garage, birthplace of Silicon Valley (Courtesy of HP)

The term *Silicon Valley* first appeared in the printed media in 1971, in a series by Don Hoefler titled *Silicon Valley in the USA*, which was published in the weekly newspaper *Electronics News*. The term was used widely from the early 1980s following the introduction of the IBM personal computer and given the high concentration of semiconductor technology companies in the area. The word *silicon* originally referred to the large number of silicon chip manufacturers in the area, as most semiconductors are made from silicon. The word *valley* refers to the Santa Clara Valley.

Bill Hewlett and Dave Packard started their new two-person company (Hewlett-Packard) in a Palo Alto garage (Fig. 7.4) on 367 Addison Avenue in 1938. Fruit orchards covered the surrounding area, as Silicon Valley as it is known today did not exist. This 12 by 18 ft garage is now a historical landmark, and it has been officially declared the *birthplace of Silicon Valley*. HP purchased the property in 2000 to preserve it for future generations.

William Shockley (one of the inventors of the transistor) moved from New Jersey to Mountain View in California to start Shockley Semiconductors in 1956. Shockley's work served as the foundation for many electronics developments. However, Shockley was a difficult person to work with and his management style soon alienated several of his employees. This led to the resignation of eight key researchers in 1957, following his decision not to continue research into silicon-based semiconductors. Shockley described them as the *traitorous eight*.

This gang of eight went on to form Fairchild Semiconductors and other companies in the Silicon Valley area in the following years. They included Gordon Moore and Robert Noyce, who founded Intel in 1968. Other employees from Fairchild Semiconductors formed companies such as National Semiconductors and Advanced Micro Devices in the Silicon Valley area in later years. Shockley Semiconductors and these new companies formed the nucleus of what became Silicon Valley.

Stanford University played an important role in the development of Silicon Valley, and Frederick Terman, the Dean of Engineering and provost of Stanford

University in the 1950s, encouraged graduates to form companies in the Silicon Valley area. Stanford University set up an industrial park (Stanford Research Park) for high-tech companies. Terman has been described as the father of Silicon Valley.

7.5 Review Questions

1. What is an integrated circuit?
2. Explain the significance of Moore's law and its relevance to the computing power of electronic devices.
3. What are the benefits of the integrated circuit?
4. Describe the early computers that were based on the integrated circuit.
5. Describe how Silicon Valley was formed.
6. Describe the role played by Stanford University in the success of Silicon Valley.

7.6 Summary

An integrated circuit consists of a set of electronic circuits on a small chip of semiconductor material, and it is much smaller than a circuit made out of independent components. Its invention placed the previously separated transistors, resistors, capacitors and wiring circuitry onto a single chip made of silicon or germanium. The integrated circuit was a revolution in computing, and it shrunk the size and cost of making electronics. It had a major influence on later developments in the computing field.

There are several generations of integrated circuits from the small-scale integration of the early 1960s, to medium-scale integration of the late 1960s, to large-scale integration (LSI) of the mid-1970s, to very large-scale and ultra large-scale integration of the 1980s. The number of transistors on a silicon chip has grown from less than 30 in the early 1960s to over a billion today.

Gordon Moore formulated *Moore's law* in 1965, in which he predicted exponential growth in transistor density. His prediction has proved to be accurate over the last 50 years, and the capabilities of many digital electronic devices are linked to Moore's law.

The earliest computers to use integrated circuits appeared in the 1960s, and their use was mainly in embedded systems. They played an important role in early aerospace projects such as the Apollo Guidance Computer and Minuteman missile. DEC's popular PDP-8 was one of the early computers to use integrated circuits, and it was introduced in 1965.

Silicon Valley today is home to the world's largest high-tech companies and thousands of start-ups. It employs hundreds of thousands of IT workers. The garage where HP was formed is considered the birthplace of Silicon Valley.

The IBM System/360

8

Abstract

The IBM System/360 was a family of mainframe computers designed and developed by IBM. It had a major impact on technology and on the computing field, and it set IBM on the road to dominate the computing field for the next 20 years, up to the arrival of personal computers in the 1980s. The user could start with a low specification member of the family and upgrade over time to a more powerful member of the family. It was the start of an era of computer compatibility, and it set IBM on the road to dominate the computing field. It was a massive $5 billion gamble by IBM, and it moved the company from its existing product lines to the unknown world of the System/360.

Key Topics

System/360
Family of computers
Gene Amdahl
Fred Brooks
The Mythical Man Month

G. O'Regan, *Introduction to the History of Computing*, Undergraduate Topics in Computer Science, DOI 10.1007/978-3-319-33138-6_8

8.1 Introduction

The IBM System/360[1] was a family of mainframe computers designed and developed by IBM. It had a major impact on technology and on the computing field, and it set IBM on the road to dominate the computing field for the next 20 years, up to the arrival of personal computers in the 1980s.

It was the beginning of an era of computer compatibility, where for the first time machines across a product line could work with each other. It meant that IBM customers could start off with a low specification member of the family and upgrade over time to a more powerful member of the family. This allowed the customer to choose the appropriate model to meet its current needs, and then as its needs evolved, it could upgrade to a more powerful member of the family. It was a massive $5 billion investment (*bet the business gamble*) by Thomas Watson Jr., and it moved IBM from its traditional business and product lines into the unknown with the gamble that the future would be the System/360.

Thomas Watson Jr.[2] announced the System/360 in 1964, and the revolutionary announcement changed business and the world of computing forever. The System/360 replaced all five of IBM's computer product lines with one strictly compatible family. It used a new computer architecture that employed hybrid integrated circuit technology, and it pioneered the 8-bit byte, which remains in use on every computer today.

The System/360 included a multiprogramming disc-based operating system, which was called OS/360. It included free software packages such as compilers for several programming languages, as well as packages for communication network capabilities [Pug:09].

The System/360 was an extremely successful product line for IBM, with orders rapidly exceeding forecasts. Its success vastly exceeded IBM's expectations, with over a thousand orders placed in the first 4 weeks after the announcement. The popularity of the System/360 made it difficult for IBM competitors (such as Burroughs, Honeywell and Sperry Rand) to compete against IBM in the general-purpose computer market.

Monthly rental prices ranged from under $3000 per month for the most basic system to over $100,000 per month for a large multisystem. The purchase cost ranged from $130,000 for a basic system to over $5 million for a large system. In 1989, 25 years after the announcement of the System/360, products based on the System/360 architecture and its extensions still accounted for over 50 % of IBM revenue.

---

[1] The number '360' (the number of degrees in a circle) was chosen to represent the ability of each computer to handle all types of applications.

[2] Thomas Watson Jr. later stated, 'The System/360 was the biggest, riskiest decision that I ever made, and I agonised about it for weeks, but deep down I believed that there was nothing that IBM couldn't do'.

8.2 Background to the Development of System/360

Thomas Watson Jr., the son of Thomas Watson Sr. (the first president of IBM), became president of IBM in 1952. He recognized that computers would play a key role for business in the years ahead, and he realized that the future of IBM was in the computer business and not in tabulators. It was clear to him that IBM needed to change, and he played a key role in transforming the company to become the world leader in the computer industry.

IBM was already a successful computer company in the 1950s. It introduced its first large computer (the IBM 701) based on vacuum tubes in 1952, the IBM 650 (Magnetic Drum Calculator) in 1954 and the IBM 704 data processing system computer in 1954. It had also played a key role in the development of the computers for the SAGE air defence system in the United States. IBM had become the market leader with a large growth in its revenue and earnings, and it employed over 100,000 people around the world.

However, within IBM there were concerns that the company had reached a plateau, and competitors were launching alternative products to IBM. The origins of the System/360 go back to the late 1950s and Watson's determination to transform IBM in order to position it for future success. IBM was supporting five different product lines by 1959, and it was becoming a major challenge to train staff to service and maintain software to support so many different computer products.

There were major problems with incompatibility between different hardware and software among the different computer vendors, as well as incompatibility among IBM's own products. IBM had an existing product line of several computers, each excellent in its own right, but all with incompatible architectures. It meant that customers who wished to move up from their existing small system to a larger system had to invest in a new system, new printers, new storage devices and new software (often totally rewritten for the new machine).

It was clear to Watson and other senior IBM executives that there was a need to develop a totally cohesive product line so that computers produced at different IBM facilities would be compatible with one another.

IBM set up a corporate wide task group to establish an overall IBM plan for its future products. The task group had the acronym SPREAD (System Programming, Research, Engineering and Design), and it completed its final report in the late 1961. It made a series of recommendations such as that there would be five processors spanning a 200-fold range in performance. IBM made the brave decision in 1962 to replace the company's entire product line of computers and to build a new family of compatible machines.

It would mean that code written for the smallest member of the family would be upwardly compatible with each of the processors in the family. Further, the various peripherals such as printers and storage devices would be compatible across the family. It was an incredibly brave decision, and *Fortune Magazine* later described it as *IBM's five billion dollar gamble*.

8.3 The IBM System/360

Thomas Watson announced the new System/360 to the world at a press conference in 1964 and said:

> The System/360 represents a sharp departure from concepts of the past in designing and building computers. It is the product of an international effort in IBM's laboratories and plants, and is the first time IBM has redesigned the basic internal architecture of its computers in a decade. The result will be more computer productivity at lower cost than ever before. This is the beginning of a new generation—not only of computers—but of their application in business, science and government.

The IBM System/360 (Fig. 8.1) was a family of small to large computers, and the concept of a *family of computers* was a paradigm shift away from the traditional *one-size-fits-all* philosophy of the computer industry, as up until then, every computer model was designed independently.

The family of computers ranged from minicomputers with 24 KB of memory to supercomputers for US missile defence systems. However, all these computers employed the same user instruction set, and the main difference was that for the larger computers, the more complex machine instructions were implemented with hardware, whereas the smaller machines used microcode.

The System/360 architecture allowed customers to commence with a lower-cost computer model and to then upgrade over time to a larger system to meet their evolving needs. The fact that the same instruction set was employed meant that the time and expense of rewriting software was avoided.

Fig. 8.1 IBM System/360 (Courtesy of IBM Archives)

Fig. 8.2 Gene Amdahl (Photo courtesy of Perry Kivolowitz)

Fig. 8.3 Fred Brooks (Photo courtesy of Dan Sears)

Gene Amdahl (Fig. 8.2) was the chief architect for the System/360, and Fred Brooks[3] was the project manager (Fig. 8.3). This family was introduced in 1964, and the IBM chairman, Thomas Watson Jr., called it the most important product announcement in the company's history.

[3] Fred Brooks wrote an influential paper *The Mythical Man Month* based on his experience as project manager for the System/360 project.

The IBM 360 family of small to large computers offered a choice of five processors and 19 combinations of power, speed and memory. There were 14 models in the family. It was successful in achieving strict compatibility in the family of computers, and the project introduced a number of new industry standards including 8-bit bytes.

A customer could start with a small member of the System/360 family and upgrade over time into a larger computer in the family. This helped to make computers more affordable for businesses, and it stimulated growth in computer use.

It was used extensively in the Apollo program to place man on the moon. The contribution by IBM computers and personnel was essential to the success of the project. IBM invested over $5 billion in the design and development of the S/360. However, the gamble paid off and it was a very successful product line for IBM.

Gene Amdahl was appointed an IBM fellow in 1965 in recognition of his contribution to IBM, and he was appointed director of IBM's Advanced Computing Systems (ACS) Laboratory in California and given freedom to pursue his own research projects. He later left IBM following disagreements on later computer development and he formed Amdahl Corporation, which later became a major competitor to IBM in the mainframe market.

Fred Brooks was the project manager for the System/360 project, which involved 5000 man-years of effort at IBM. Brooks recorded his experience as project manager in a famous project management book titled *The Mythical Man Month* [Brk:75]. This book which appeared in 1975 considered the challenge of delivering a major project (of which software is a key constituent) on time, on budget and with the right quality. Brooks described it as *my belated answer to Tom Watson's probing question as to why programming is hard to manage.*

For a more detailed account of the System/360 revolution, see the excellent IBM article 'The 360 Revolution' by Chuck Boyer [Boy:04]. For more detailed information on Brooks and Amdahl, see [ORg:13, ORg:15].

8.4 Review Questions

1. Why did IBM decide to retire its existing product line and develop the System/360?
2. What were the main risks in developing the System/360?
3. What were the advantages of developing the System/360?
4. What new industry standards followed from the System/360?
5. What is a family of computers?
6. Describe the contributions of Gene Amdahl to the computing field.
7. Describe the contributions of Fred Brooks to the computing field.

8.5 Summary

The IBM System/360 was a family of small to large computers, and it was a paradigm shift away from the traditional 'one-size-fits-all' philosophy of the computer industry, as up until then, every computer model was designed independently.

The family ranged from minicomputers with 24 KB of memory to supercomputers for US missile defence systems. However, all these computers employed the same user instruction set, and the main difference was that for the larger computers, the more complex machine instructions were implemented with hardware, whereas the smaller machines used microcode.

The System/360 architecture allowed customers to commence with a lower-cost computer model and to then upgrade over time to a larger system to meet their evolving needs. The fact that the same instruction set was employed meant that the time and expense of rewriting software was avoided.

Gene Amdahl was the chief architect for the System/360 and Fred Brooks was the project manager. Fred Brooks later wrote an influential project management book, which was concerned with the challenge of delivering a major project (of which software is a key part) on time, on budget and with the right quality.

Minicomputers and Later Mainframes

9

Abstract

The *minicomputer* was a new class of low-cost computers that arose during the 1960s, and its development was facilitated by the introduction of integrated circuits and their improved performance and declining cost. Minicomputers were distinguished from the large mainframe computers by price and size, and they formed a class of the smallest general-purpose computers. We discuss minicomputers such as DEC's PDP-1, PDP-11 and VAX-11/780 minicomputers, which were popular with the engineering and scientific communities. DEC became the second largest computer company in the world in the late 1980s, but it was too slow in reacting to the rise of the microprocessor and the revolution in home computers. Later mainframes are discussed including the Amdahl 470V/6 and the intense competition between IBM and Amdahl in the high-end mainframe market.

Key Topics

DEC
Minicomputers
PDP-11
VAX-11/780
Amdahl 470
IBM System/370

G. O'Regan, *Introduction to the History of Computing*, Undergraduate Topics in Computer Science, DOI 10.1007/978-3-319-33138-6_9

9.1 Introduction

The *minicomputer* was a new class of low-cost computers that arose during the 1960s. The development of minicomputers was facilitated by the introduction of integrated circuits and their improved performance and declining cost. Minicomputers were distinguished from the large mainframe computers by price and size, and they formed a class of the smallest general-purpose computers.

Mainframes were large expensive machines (typically costing over $1 million) and they required separate rooms for technicians and operation, whereas minicomputers cost well under $100,000 and they were designed for direct, personal interaction with the programmer.

Digital Equipment Corporation (DEC) and Control Data Corporation (CDC) introduced small or minicomputers in the early 1960s. These included DEC's PDP-1, which was released in 1961, and the CDC-160A, which was released in 1960. These machines cost $110,000 and $60,000, respectively, which was a fraction of the cost of a mainframe computer.

The DEC PDP series of minicomputers became popular in the 1960s. The PDP-8 minicomputer (Fig. 7.3) was released in 1965, and it was a 12-bit machine with a small instruction set. The PDP-11 was a highly successful series of 16-bit minicomputers, and it remained a popular product for over 20 years from its release in 1970 to the early 1990s.

Gene Amdahl was the chief architect for the IBM System/360, and he resigned from IBM to set up Amdahl Corporation in 1970. His goals were to develop a mainframe that would provide better performance than the existing IBM machines, and do so at a lower cost, as well as being compatible with IBM hardware and software.

Amdahl Corporation launched its first product, the Amdahl 470V/6, in 1975. This was an IBM S/370 compatible mainframe that could run IBM software, and so it was an alternative to a full IBM proprietary solution. It meant that companies around the world now had the choice of continuing to run their software on IBM machines or purchasing the cheaper and more powerful IBM compatibles produced by Amdahl. Amdahl Corporation became a major competitor to IBM in large-scale computer placements.

Amdahl Corporation's success led to a price war with IBM, with the latter offering discounts to its customers to protect its market share.

9.2 DEC's Minicomputers

Ken Olsen and Harlan Anderson founded Digital Equipment Corporation (DEC) in 1957. It was a spin-off from MIT's Lincoln computer laboratory, and it was an innovative and forward-thinking company. It became the second largest computer company in the world in the late 1980s, with revenues of over $14 billion and over 100,000 employees. It dominated the minicomputer era from the 1960s to the 1980s, with its PDP and VAX series of computers, which were very popular with the engineering and scientific communities.

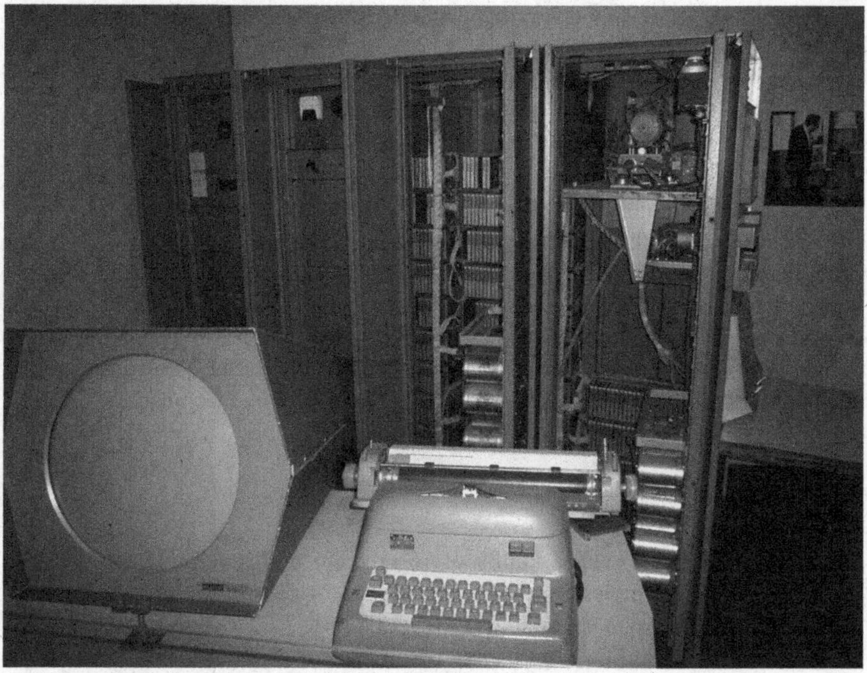

Fig. 9.1 The PDP-1 computer

DEC's first computer, the *Programmed Data Processor* (PDP-1), was released in 1961 (Fig. 9.1). This 18-bit machine was a relatively inexpensive computer for the time, and it cost $110,000. The existing IBM mainframes cost over $2 million, and so DEC's minicomputers were relatively affordable to businesses. It was a simple and reasonably easy-to-use computer with 4000 words of memory.

The PDP series of minicomputers were elegant and reasonably priced and dominated the new minicomputer market segment. They were an alternative to the multimillion dollar mainframe computers offered by IBM to large corporate customers. Research laboratories, engineering companies and other organizations with large computing needs all used DEC's minicomputers

The PDP-8 minicomputer (Fig. 7.3) was released in 1965, and it was a 12-bit machine with a small instruction set. It was a major commercial success for DEC with many sold to schools and universities. The PDP-11 was a highly successful series of 16-bit minicomputer, and it remained a popular product for over 20 years from the 1970s to the 1990s.

Gordon Bell was one of the earliest employees of the company, and he played an important role in the development of the PDP family of minicomputers. He designed the multiplier/divider unit and the interrupt system for the PDP-1 computer, which built upon work done at the MIT Lincoln Laboratory. He later became vice president of research and development at DEC, and he was the architect of several PDP computers. He later led the development of the 32-bit VAX series of computers, and he was involved in the design of around 30 microprocessors.

The VAX series of minicomputers were derived from the best-selling PDP-11, and the VAX was the first widely used 32-bit minicomputer. The VAX-11/780 was released in 1978, and it was a major success for the company. The VAX product line was a competitor to the IBM System/370 series of mainframe computers. The VAX minicomputers used the Virtual Memory System (VMS) operating system.

The rise of the microprocessor and microcomputer led to the availability of low-cost personal computers, and this later challenged DEC's product line. DEC was slow in recognizing the importance of these developments, and Olsen's statement from the mid-1970s 'There is no need for any individual to have a computer in his home' suggests that DEC were totally unprepared for the revolution in personal computing and its threat to DEC's business. DEC was too late in responding to the paradigm shift in the industry, and this proved to be fatal for the company. Compaq acquired DEC in 1998 for $9.8 billion, and HP later acquired Compaq.

9.2.1 PDP-11

The PDP-11 (Fig. 9.2) was a family of 16-bit minicomputers produced by DEC from 1970 up to the early 1990s. It was designed by Harold McFarland, with the prototype ready in 1969 and the PDP/11 released in 1970. There were several models in the PDP-11 family.

It was one of DEC's most successful computers, with over 600,000 machines sold. It was the only 16-bit computer made by the company, as its successor was the 32-bit VAX-11 series. It started its life as a minicomputer and ended its life as macro/super-microcomputer. The release price of the PDP-11 in 1970 was a very affordable $20,000.

Its central processing unit had eight 16-bit registers, six general-purpose registers, the stack pointer and a program counter. It included software such as an editor, debugger and utilities. The size of its memory was 128 KB.

The PDP-11 was very useful for multi-user and multitask applications, and the first version of the UNIX operating system ran on a PDP-11/20 in 1970. The VAX line at Digital began as an enhancement to the PDP-11 architecture.

9.2.2 The VAX-11/780

The Virtual Address eXtension (VAX) was a family of minicomputers produced by DEC from the mid-1970s up to the late 1980s. This family used processors implementing the VAX instruction set architecture, and its members included minicomputers such as the VAX-11/780, /782, /784, /785, /787, /788, /750, /725 and /730. The VAX product line was a competitor to the IBM System/370 series of computers.

The VAX series was derived from the PDP-11 minicomputer and the VAX-11/780 (Fig. 9.3) was the first member of the family. It was the first widely used 32-bit minicomputer, and it was released in 1978. It was the first one MIPS (million instructions per second) machine, and it was a major success for the company.

Fig. 9.2 PDP-11

Several programming languages including Fortran 77, BASIC, COBOL and Pascal were available for the machine. The VAX-11/780 used the DEC VMS operating system, which was a multi-user, multitasking and virtual memory operating system. The VAX-11/780 remained the base system that every computer benchmarked its speed against for many years.

It supported 128KB–8MB of memory through one or two memory controllers, and the memory was protected with error-correcting codes. Each memory controller could support 128KB–4MB of memory. For more detailed information on DEC, see [Sch:04].

9.3 The War Between IBM and Amdahl

Gene Amdahl (Fig. 8.2) resigned from IBM to set up Amdahl Corporation in 1970, and his goal was to develop a mainframe that would be compatible with the IBM System/360. Further, he intended that it would provide a better performance at a lower cost than the existing IBM machine. Amdahl revised his plans to launch an IBM-compatible System/360 mainframe following IBM's introduction of its IBM System/370 mainframe.

Amdahl Corporation launched its first product, the Amdahl 470V/6, in 1975. This was an IBM System/370 compatible mainframe that could run IBM software, and so it was an alternative to a full IBM proprietary solution. It meant that companies around the world now had the choice of continuing to run their software on IBM machines or purchasing the cheaper and more powerful IBM compatibles produced by Amdahl.

Fig. 9.3 VAX-11/780

Amdahl's first customer was the NASA Goddard Institute for Space Studies, which was based in New York. The Institute needed a powerful computer to track data from its Nimbus weather satellite, and it had a choice between a well-established company such as IBM and an unknown company such as Amdahl. It seemed likely that IBM would be the chosen supplier. However, the institute was highly impressed with the performance of the Amdahl 470V/6, and its cost was significantly less than the IBM machine.

The Amdahl 470 competed directly against the IBM System/370 family of mainframes. It was compatible with IBM hardware and software but cheaper than the IBM product: i.e. the Amdahl machines provided better performance for less money. Further, the machine was much smaller than the IBM machine due to the use of large-scale integration (LSI) with many integrated circuits on each chip. This meant that the Amdahl 470 was one-third of the size of IBM's 370. It was over twice as fast and sold for about 10 % less than the IBM 370.

IBM's machines were water-cooled, while Amdahl's were air-cooled, which decreased installation costs significantly. Machine sales were slow initially due to concerns over Amdahl Corporation's long-term survival and the risks of dealing with a new player. IBM had a long-established reputation as the leader in the computer field. The University of Michigan was Amdahl's second customer, and it used the 470 in its education centre. Texas A&M University was Amdahl's third customer, and they used the 470 for educational and administrative purposes. Amdahl Corporation was well on its way to success, and by 1977 it had over 50 470V/6 machines installed at various customer sites.

IBM launched a new product, the IBM 3033, in 1977 to compete with the Amdahl 470. However, Amdahl Corporation responded with a new machine, the 470V/7, which was one and a half times faster than the 3033 and only slightly more expensive. Customers voted with their feet and chose Amdahl as their supplier, and by late 1978, it had sold over a hundred of the 470V/7 machines.

IBM introduced a medium-sized computer, the 4300 series, in early 1979, and in late 1980, it announced plans for the 3081 processor which would have twice the performance of the existing 3033 on its completion in late 1981. In response, Amdahl announced the 580 series (Fig. 9.4), which would have twice the performance of the existing 470 series. The 580 series was released in mid-1982, but their early processors had some reliability problems and lacked some of the features of the new IBM product.

Amdahl moved into large system multiprocessor design from the mid-1980s. It introduced its 5890 model in late 1985 and its superior performance allowed Amdahl to gain market share and increase its sales to approximately $1 billion in 1986. It now had over 1300 customers in around 20 countries around the world. It launched a new product line, the 5990 processor, in 1988, and this processor outperformed IBM by 50 %. Customers voted with their feet and chose Amdahl as their supplier.

Fig. 9.4 Amdahl 5860 (Courtesy of Robert Broughton, University of Newcastle)

It was clear that Amdahl was now a major threat to IBM in the high-end mainframe market. Amdahl had a 24 % market share and annual revenues of $2 billion at the end of 1988. This led to a price war with IBM, with the latter offering discounts to its customers to protect its market share. Amdahl responded with its own discounts, and this led to a reduction in profitability for the company.

The IBM personal computer was introduced in the early 1980s, and by the early 1990s, it was clear that the major threat to Amdahl was the declining mainframe market. Revenue and profitability fell, and Amdahl shut factory lines and cut staff numbers. By the late 1990s, Amdahl was making major losses, and there were concerns about the future viability of the company.

It was clear by 2001 that Amdahl could no longer effectively compete against IBM following IBM's introduction of its 64-bit zSeries architecture. Amdahl had invested a significant amount in research on a 64-bit architecture to compete against the zSeries, but the company estimated that it would take a further $1 billion and two more years to create an IBM-compatible 64-bit system. Further, it would be several years before they would gain any benefit from this investment as there were declining sales in the mainframe market due to the popularity of personal computers.

By late 2001, the sales of mainframes accounted for just 10 % of Amdahl's revenue, with the company gaining significant revenue from the sale of Sun servers. Amdahl became a wholly owned subsidiary of Fujitsu in 1997, and it exited the mainframe business in 2002. Today, it focuses on the server and storage side as well as on services and consulting.

For more detailed information on Gene Amdahl, Amdahl Corporation, IBM and Digital Equipment Corporation, see [ORg:13, ORg:15].

9.4 Review Questions

1. What is a minicomputer?
2. What factors led to the introduction of the minicomputer?
3. Describe the achievements of Gene Amdahl.
4. Describe the competition between Amdahl Corporation and IBM in the mainframe market.
5. What factors led to the demise of DEC and Amdahl?
6. What could have DEC and Amdahl done differently?
7. Describe the achievements of Gordon Bell.

9.5 Summary

The minicomputer was a new class of low-cost computers that arose during the 1960s. The development of minicomputers was facilitated by the introduction of integrated circuits, as this helped to reduce cost and size of computers. Minicomputers were distinguished from the large mainframe computers by price and size, and they formed a class of the smallest general-purpose computers.

Minicomputers typically cost well under $100,000 and so were relatively inexpensive compared to mainframes. They were designed for direct, personal interaction with the programmer. DEC, CDC and HP introduced small or minicomputers in the early 1960s.

The PDP-11 was a highly successful series of 16-bit minicomputers, and it remained a popular product from the 1970s to the 1990s. The VAX series of minicomputers were derived from the PDP-11, and it was the first widely used 32-bit minicomputer. The VAX-11/780 was released in 1978, and it was a major success for DEC.

The rise of the microprocessor and microcomputer led to the availability of low-cost home and personal computers, and this paradigm shift later challenged the mainframe and minicomputer market. DEC was too late in responding to the paradigm shift in the industry, and this proved to be fatal for the company.

Gene Amdahl resigned from IBM to set up Amdahl Corporation in 1970, and his goals were to develop a mainframe that would be compatible with the IBM System/360 and that would provide better performance at a lower cost than the IBM machine. Amdahl Corporation launched its first product, the Amdahl 470V/6, in 1975, and this computer was compatible with the IBM System/370 mainframe. It meant that companies now had the choice of continuing to run their software on IBM machines or purchasing the cheaper and more powerful IBM compatibles produced by Amdahl.

Amdahl became a major threat to IBM in the high-end mainframe market, as customers placed orders with Amdahl at IBM's expense. By the late 1980s, it had 24 % market share and annual revenues of $2 billion. However, as the mainframe market declined in the 1990s, Amdahl failed to adapt to the rise of the personal computer, and it went through major financial difficulties and was taken over by Fujitsu.

The Microprocessor Revolution 10

Abstract

A microprocessor is a central part of a modern personal computer (or computer device). It integrates the functions of a central processing unit (the part of a computer that processes the program instructions) onto a single integrated circuit and places a vast amount of processing power in a tiny space.

Intel's invention of the microprocessor in 1971 was a revolution in computing, and it placed the power of a computer on a tiny chip. It was initially developed as an enhancement to allow users to add more memory to their units. However, it soon became clear that the microprocessor had great potential for everything from calculators to cash registers and traffic lights. Its invention made personal computers, tablets and mobile phones possible.

We discuss early microprocessors such as the Intel 4004, the 8-bit Intel 8080 and the 8-bit Motorola 6800. The 16-bit Intel 8086 was introduced in 1978 and the 16/32-bit Motorola 68000 was released in 1979. The 8-bit Intel 8088 (the cheaper 8-bit variant of the Intel 8086) was introduced in 1979, and it was chosen as the microprocessor for the IBM personal computer.

Key Topics

Microprocessor
Intel 4004
Intel 8008
Intel 8080
Intel 8088
Motorola 68000

G. O'Regan, *Introduction to the History of Computing*, Undergraduate Topics in Computer Science, DOI 10.1007/978-3-319-33138-6_10

10.1 Introduction

A *microprocessor* is a central part of a modern personal computer (or computer device). It integrates the functions of a central processing unit (the part of a computer that processes the program instructions) onto a single integrated circuit and places a vast amount of processing power in a tiny space.

Intel's invention of the microprocessor in 1971 was a revolution in computing, and it placed the power of a computer on a tiny chip. It was initially developed as an enhancement to allow users to add more memory to their units. However, it soon became clear that the microprocessor had great potential for everything from calculators to cash registers and traffic lights. Its invention made personal computers, tablets and mobile phones possible.

Computers in the 1960s were large and expensive, and they typically filled an entire room. They were available only to a small number of individuals and government laboratories. The invention of the transistor by Shockley and others at Bell Labs had helped to reduce the size and cost of a computer.

The later invention of the integrated circuit by Jack Kilby of Texas Instruments, and improved upon by Robert Noyce and others at Fairchild Semiconductors, meant that several transistors could now be placed on a chip, leading to further reductions in the size and cost of machines. However, large-scale integration where a large number of transistors could be placed onto a silicon chip was still a long way away.

Several employees left Fairchild Semiconductors in the late 1960s to form their own semiconductor companies in the Silicon Valley area. They formed companies such as Intel, National Semiconductors and Advanced Micro Devices (AMD). Intel began operations making memory chips and it delivered its first product the 64-bit SRAM chip (the 3101) to Honeywell in 1969. It introduced a DRAM chip (the 1103) in 1970, and in 1971, it introduced the microprocessor, an invention that transformed the computing field.

10.2 Invention of the Microprocessor

The invention of the microprocessor (initially called microcomputer) in 1971 was a revolution in computing, with the power of a computer now available on a tiny microprocessor chip.

The microprocessor is essentially a computer on a chip, and its invention made hand-held calculators and personal computers (PCs) possible. Intel's microprocessors are used on the majority of personal computers and laptops around the world.

The invention of the microprocessor happened by accident rather than design. The Nippon Calculating Machine Corporation (later known as Busicom), a Japanese company, requested Intel to design a set of integrated circuits for its new family of high-performance programmable calculators. At that time, it was standard practice to custom design all logic chips for each customer's product, and this clearly limited the applicability of a logic chip to a specialized domain.

The design proposed by Busicom required 12 integrated circuits. Ted Hoff, an Intel engineer, studied Busicom's design and he rejected it as unwieldy. He proposed a more elegant solution requiring just four integrated circuits, and his design included a chip that was a general-purpose logic device (microprocessor) that derived its application instructions from the semiconductor memory. Busicom accepted his proposed design, and Intel engineers then implemented it.

Hoff's 4004 microprocessor design included a central processing unit (CPU) on one chip. It contained 2300 transistors on a one-eighth by one-sixth inch chip surrounded by three ICs containing ROM, shift registers, input/output ports and RAM.

Busicom had exclusive rights to the design and components, but following discussion and negotiations, Busicom agreed to give up its exclusive rights to the chips. Intel shortly afterwards announced the availability of the first microprocessor, the Intel 4004 (Fig. 10.1).

This was the world's first microprocessor, and although it was initially developed as an enhancement to allow users to add more memory to their units, it soon became clear that the microprocessor could be applied to many other areas.

This small Intel 4004 microprocessor chip was launched in late 1971, and it could execute 60,000 operations per second. The tiny chip had an equivalent computing power as the large ENIAC which used 18,000 vacuum tubes and took up the space of an entire room [ORg:11].

The Intel 4004 sold for $200 and for the first time affordable computing power was available to designers of all types of products. The introduction of the microprocessor was a revolution in computing, and its invention had applications to everything from traffic lights to medical instruments and to the development of home and personal computers.

Fig. 10.1 Intel 4004 microprocessor

Gary Kildall was one of the early people to recognize the potential of the microprocessor as a computer in its own right, and he began writing experimental programs for the Intel 4004 microprocessor in the early 1970s. Kildall worked as a consultant with Intel on the later 8008 and 8080 microprocessors.

He developed the first high-level programming language for a microprocessor (PL/M) in 1973, which enabled programmers to write applications for microprocessors. He developed the CP/M operating system (Control Program for Microcomputers) in the same year. CP/M allowed the Intel 8080 microprocessor to control a floppy disk drive allowing files to be read and written to and from an 8 inch floppy disk. CP/M made it possible for computer hobbyists and companies to build the first home computers.

Kildall made CP/M hardware independent by creating a separate module called the BIOS (basic input/output system). He added several utilities such as an editor, debugger and assembler, and by 1977, several manufacturers included CP/M with their systems. He set up Digital Research Inc. (DRI) in 1976 to develop, market and sell the CP/M operating system.

10.3 Early Microprocessors

Intel has developed more and more powerful microprocessors since its introduction of the Intel 4004. The Intel 8008 was launched in 1972, and this was a reasonably successful product. It led to the 8-bit Intel 8080 microprocessor, which was released in 1974. The Intel 8080 was the first general-purpose microprocessor, and it was sold for $360: i.e. a whole computer on one chip was sold for $360, while conventional computers sold for thousands of dollars. The Intel 8080 soon became the industry standard, and Intel became the industry leader in the 8-bit market. The 8080 played an important role in starting personal computer development, as it attracted the interest of computer developers and engineers.

Motorola introduced its first microprocessor, the 8-bit 6800 microprocessors (Fig. 10.2), in 1974, and this microprocessor was used in automotive, computing and video games. It contained over 4000 transistors. It competed against the Intel 8080 microprocessor, and it was used in some early home computer kits.

Fig. 10.2 Motorola 6800 microprocessor

National Semiconductors introduced its 16-bit IMP-16 in 1973 and an 8-bit version, the IMP-8, in 1974. Texas Instruments introduced the first single-chip microprocessor, the PACE, in 1974, and it introduced its first 16-bit microprocessor, the TMS 9900, in 1976. MOS Technology introduced its 8-bit 6502 in 1975, and Zilog introduced its Z80 in 1976.

The 16-bit Intel 8086 was introduced in 1978, but it soon faced competition from Motorola, which introduced its 16/32-bit 68000 microprocessor in 1979. The Intel 8088 is an 8-bit variant of the 8086, and it was introduced in 1979. The Motorola 68000 was a hybrid 16/32-bit microprocessor that had a 16-bit data bus, but it could perform 32-bit calculations internally. It was used on various Apple Macintosh computers, the Atari ST and the Commodore Amiga.

The first single-chip 32-bit microprocessor was AT&T Bell Labs BELLMAC-32A, which was introduced in 1982. Motorola introduced its 32-bit 68020 microprocessor in 1984, and this microprocessor contained 200,000 transistors on a three-eighths inch square chip.

IBM considered several microprocessors for its IBM PC including the IBM 801 processor, the Motorola 68000 microprocessor and the Intel 8088 microprocessor. IBM chose the Intel 8088 chip (which was cheaper than the 16-bit Intel 8086), and it took a 20 % stake in Intel leading to strong ties between both companies.

Today, Intel's microprocessors are used on most personal computers around the world, and the contract to supply the Intel 8088 microprocessor was a major turning point for the company. Intel had been focused more on the sale of dynamic random access memory chips, with sales of microprocessors in thousands or in tens of thousands. However, sales of microprocessors rocketed following the introduction of the IBM PC, and soon sales were in tens of millions of units.

The introduction of the IBM PC was a revolution in computing, and there are hundreds of millions of computers in use around the world today. It placed computing power in the hands of ordinary users, and today's personal computers are more powerful than the mainframes that were used to send man to the moon. The cost of computing processing power has fallen exponentially since the introduction of the first microprocessor, and Intel has played a key role in squeezing more and more transistors onto a chip leading to more and more powerful microprocessors and personal computers.

10.4 A Selection of Semiconductor Companies

Robert Noyce and Gordon Moore founded Intel (Integrated Electronics) in 1968. Today, it is an American semiconductor giant with headquarters at Santa Clara in California. It is one of the largest semiconductor manufacturers in the world, with plants in the United States, Europe and Asia. It has played an important role in shaping the computing field with its invention of the microprocessor in 1971. It is the inventor of the x86 series of microprocessors that are used in most personal computers, and the company is renowned for its leadership in the microprocessor industry and for its excellence and innovation in microprocessor design and manufacturing.

Noyce and Moore left Fairchild Semiconductors to set up Intel, and the initial focus of the company was on semiconductor memory products and to make semiconductor memory practical. Its goal was to create large-scale integrated (LSI) semiconductor memory, and it introduced a number of products including the Intel 1103, which was a one-kilobit (KB) dynamic random access memory (DRAM) integrated circuit.

Motorola set up a research lab in 1952 to take advantage of the potential of semiconductors, and by 1961 it was mass-producing semiconductors at a low cost. It introduced a transistorized walkie-talkie in 1962 as well as transistors for its Quasar televisions. Its microprocessors have played an important role in the computing field. These include the influential 68000 and PowerPC architecture, which were used in the Apple Macintosh and Power Macintosh personal computers. Motorola's semiconductor business was spun off to become a separate company called Freescale Semiconductor Inc. in 2004.

Advanced Micro Devices was formed by Jerry Sanders and several of his colleagues from Fairchild Semiconductors in 1969. It initially acted as a second-source supplier of microchips designed by Fairchild and National Semiconductors, and it later acted as second supplier for the x86 chips produced by Intel. AMD produces microprocessors, motherboards and chipsets, and it is the second largest supplier of x86-based microprocessors.

National Semiconductors was founded in Connecticut by Bernard Rothlein and several of his colleagues from Sperry Rand Corporation. It introduced the 16-bit IMP-16 microprocessor in 1973 and the 8-bit version, the IMP-8, in 1974. National Semiconductors was taken over by Texas Instruments in 2011.

Texas Instruments (TI) is an American electronics company that was formed in 1951, and its headquarters are in Dallas. It is one of the largest manufacturers of semiconductors in the world, and it produces a wide range of semiconductor products including chips for mobile phones, calculators, microcontrollers, digital signal processors, analog semiconductors and multicore processors.

It commenced research on transistors in the early 1950s, and it introduced one of the first transistor radios in 1954. It invented the integrated circuit in 1958; PACE, the first single-chip microprocessor, was introduced in 1974; and the TMS 9900, its first 16-bit microprocessor, was released in 1976.

MOS Technology was formed in 1969 initially as a second supplier of calculator chips for Texas Instruments. Several Motorola designers of the Motorola 6800 microprocessor joined the company in 1975, and their knowledge allowed MOS to develop the 6501 and 6502 microprocessors. MOS Technology was taken over by Commodore in 1976.

10.5 Review Questions

1. What is a microprocessor?
2. What is the significance of the Intel 4004?
3. Why is the invention of the microprocessor considered a revolution in computing?
4. What are the main contributions made by Motorola to the semiconductor field?
5. Why did so many employees leave Fairchild Semiconductors to set up companies in Silicon Valley? What companies did they form?
6. What are the main contributions made by Intel to the semiconductor field?
7. Explain the significance of PL/M and CP/M?

10.6 Summary

A microprocessor is a central part of a modern personal computer (or computer device), and it places a vast amount of processing power on a tiny chip. Intel's invention of the microprocessor in 1971 changed computing forever, and it placed the power of a computer on a tiny chip.

The microprocessor was initially developed as an enhancement to allow users to add more memory to their units. However, it soon became clear that the microprocessor had applications to many other areas. Its invention led to personal computers, tablets and mobile phones.

The invention of the microprocessor happened by accident rather than design, and it was initially developed as part of the design to allow users to add more memory to their units. The design solution included a general-purpose chip that derived its application instructions from the semiconductor memory. This was the Intel 4004 microprocessor.

Home Computers

11

Abstract
We consider a selection of home and personal computers, including early home computers such as the MITS Altair 8800, which was introduced in early 1975; the Apple I and II computers, which were released in 1976 and 1977, respectively; the Commodore PET computer, which was introduced in 1977; the Atari 400 and 800 computers, which were released in 1979; the popular Commodore 64 computer, which was introduced in 1982; and the Sinclair ZX 81 and ZX Spectrum computers, which were released in 1980 and 1981, respectively. We discuss later Atari and Amiga computers and the Apple Macintosh computer, which was a major milestone in computing.

Key Topics
Xerox Alto
MITS Altair 8800
Apple I and II computers
Atari 400 and 800
Commodore PET
Amiga
Commodore 64
Sinclair ZX Spectrum
Apple Macintosh

G. O'Regan, *Introduction to the History of Computing*, Undergraduate Topics in Computer Science, DOI 10.1007/978-3-319-33138-6_11

11.1 Introduction

The invention of the microprocessor was a revolution in computing, and it led to the development of home and personal computers. We consider a selection of home and personal computers in this chapter, including early home computers such as the MITS Altair 8800, the Apple I and II computers, the Commodore PET computer, the Atari 400 and 800 computers and the Commodore 64 computer. We discuss later Atari and Amiga computers and the Apple Macintosh computer, which was a major milestone in computing. We will discuss the introduction of the IBM personal computer in Chap. 12.

Many of the early home computers discussed in this chapter were based on the 8-bit MOS 6502 microprocessor. The MITS Altair 8800 is an exception, as it was based on the Intel 8080 microprocessor, and it was one of the earliest home computers when it was introduced in late 1974.

Later home and personal computers used a variety of microprocessors. The ZX Spectrum home computer was based on the 8-bit Zilog Z80 microprocessor; the Apple Macintosh was based on the Motorola 68000 microprocessor, as was the Amiga 1000. The Atari personal computer was based on the Intel 8088 microprocessor.

We start with a discussion of the Xerox Alto computer, which was developed at Xerox PARC. This computer pioneered several key concepts in personal computing, and it had a major impact on the design of the Apple Macintosh.

11.2 Xerox Alto Personal Computer

The Xerox Alto (Fig. 11.1) was one of the earliest personal computers, and it was introduced in early 1973. Chuck Thacker and others at Xerox designed it, and it was one of the first computers to use a mouse-driven graphical user interface. It was designed for individual rather than home use, and a single person sitting at a desk used it. It was essentially a small minicomputer rather than a personal computer, and it was unlike modern personal computers in that it was not based on the microprocessor. The significance of the Xerox Alto is that it had a major impact on the design of early personal computers and especially on the design of the Apple Macintosh computer.

Butler Lampson wrote a famous memo to the management in Xerox in 1972 [Lam:72], in which he requested funds to construct a number of Alto workstations. He made the case for the development of the Alto, and he outlined his vision of personal computing in the memo. His vision described the broad range of applications to which the Xerox Alto could be applied.

He outlined a vision of distributed computing, where several Xerox Alto workstations would form a network of computers, with each computer user having his own files and communicating with other users to interchange or share information. He argued that the development of the Alto would allow the theory that cheap

Fig. 11.1 Xerox Alto

personal computers would be extremely useful to be tested and demonstrated comprehensively to be the case.

This memo led to the development of a network of Alto in the mid-1970s, the development of Ethernet technology for connecting computers in a network, the development of a mouse-driven graphical user interface, the development of a WYSIWYG editor and laser printing and the development of the Smalltalk object-oriented programming language.

The cost of the Alto machine was approximately $10,000, and this was significantly less than the existing mainframes and minicomputers. The machine was capable of performing almost any computation that a DEC PDP-10 machine could perform. For a more detailed account of the contributions of Xerox PARC to the computing field, see [Hil:00].

11.3 MITS Altair 8800

Micro Instrumentation and Telemetry Systems (MITS) was founded by Ed Roberts and others in 1969. Roberts had a background in electronics from the US military, and the company began in Robert's garage in New Mexico. Its initial focus was to design and sell electronic kits to model rocket enthusiasts, which had become a popular hobby in the 1960s, due to manned space flights and the race to the moon.

The next product that the company introduced was the MITS 816 calculator kit, which included six LSI integrated circuits designed to make a calculator with the four basic arithmetic functions. The calculator kit featured on the November 1971 cover of *Popular Electronics*, which was a popular American electronics magazine that appeared from the mid-1950s to the late 1990s.

MITS began working on the Altair 8800 home computer (Fig. 11.2) in 1974, and the prototype was available in October of that year. The cover page of the January 1975 edition of Popular Electronics featured an early design of the Altair 8800, and this publicity helped in generating sales that vastly exceeded expectations. Over 5000 machines were delivered by August 1975, and the home computer kit version (which was assembled by the customer) cost $439, whereas the fully assembled version cost $621.

The home kit included assembly instructions, a metal case, a front panel with switches, a power supply, a motherboard with expansion slots, various cards to plug into the expansion slots, as well as any other components required to build the computer. The actual assembly was quite a challenge as it involved careful soldering and assembly. There was no actual keyboard or monitor, which meant that the task of programming the machine was non-trivial and required the user to program in machine language and watch the LEDs on the front panel to get the results. Several expansion cards (e.g. for keyboard, monitor and data storage) were soon released, and this made it easier to use. The Altair 8800 used the 8-bit Intel 8080 microprocessor, which was introduced in 1974.

Bill Gates and Paul Allen developed a BASIC interpreter for the Altair 8800, and the 4k/8k versions of BASIC were released in July 1975. It cost the customer an additional $60/$75 when purchasing an Altair 8800. Gates and Allen formed Microsoft later in 1975, and Altair BASIC was their first product.

Fig. 11.2 MITS Altair computer (Photo public domain)

11.4 Apple I and II Home Computers

Steve Jobs and Steve Wozniak formed Apple Computer, Inc. in 1976, and the company commenced operations in Job's family garage. Their goal was to develop a user-friendly alternative to the existing mainframe and minicomputers produced by IBM and Digital. Wozniak was responsible for product development and Jobs for marketing. Jobs and Wozniak were both college dropouts, and both attended the Homebrew Computer Club of computer enthusiasts in Silicon Valley during the mid-1970s.

The Apple I computer was released in 1976, and it retailed for $666.66. It generated over $700,000 in revenue for the company, but it was mainly of interest to computer hobbyists and engineers. This was due to the fact that it was not a fully assembled personal computer as such, and it was essentially an assembled motherboard that lacked features such as a keyboard, monitor and case. It used a television as the display system, and it had a cassette interface to allow programs to be loaded and saved. It used the inexpensive MOS Technology 6502 microprocessor chip, which had been released earlier that year, and Wozniak had already written a BASIC interpreter for this chip.

The Apple II computer (Fig. 11.3) was released in 1977, and it was a significant advance on its predecessor. It was a personal computer with a monitor, keyboard and case, and it was one of the earliest computers to come preassembled. It was a popular 8-bit home computer, and it was one of the earliest computers to have a colour display with colour graphics.

The BASIC programming language was built-in, and it contained 4 K of RAM (which could be expanded to 48 K). The VisiCalc spreadsheet program was released on the Apple II, and this helped to transform the computer into a credible business

Fig. 11.3 Apple II computer (Photo public domain)

machine. The Apple II retailed for $1299, and it was a major commercial success for Apple generating over $139 million in revenue for the company. For more detailed information on Apple, see [ORg:15].

11.5 Commodore PET

Commodore Business Machines was a leading North American home computer and electronics manufacturing company. It played an important role in the development of the home computer industry in the 1970s and 1980s, and it is especially famous for its development of the Commodore PET computer, which was very popular in the education field. It also developed the VIC-20 and Commodore 64 home computers, which were very popular machines.

Commodore initially manufactured typewriters for the North American market, and it diversified into the manufacture of mechanical calculators from the early 1960s. It introduced both consumer and scientific calculators in the late 1960s, and by the early 1970s, it was one of the most popular brands for calculators. The calculators used Texas Instruments chips, but when Texas Instruments entered the calculator market in the mid-1970s, Commodore was unable to compete with the prices offered by Texas.

Commodore purchased the semiconductor company, MOS Technology, with the intention of using MOS chips in its calculators. However, Chuck Peddle, one of MOS's employees, convinced Commodore that the future was in computers and not calculators. Commodore used one of MOS Technology's chips, the 8-bit 6502, to enter the home computer market in 1977 with the launch of its Commodore Personal Electronic Transactor (PET) computer.

This Commodore PET was very popular in the education market, and one of its models was called the *Teacher's PET*. It used the MOS 8-bit 6502 microprocessor, which was designed by Chuck Peddle and others at MOS Technology. The 6502 controlled the screen, keyboard, the cassette recorder and any peripherals connected to the expansion ports. The machine used the Commodore BASIC operating system. There were several models of the Commodore PET introduced during its lifetime including the PET 2001 series, the PET 4000 series and the SuperPET 8000 series.

The first model introduced was the PET 2001 (Fig. 11.4), which had either 4 Kb or 8 Kb of RAM. It had a built-in monochrome monitor with 40 × 25 character graphics enclosed in a metal case. It included a magnetic data storage device known as a datasette (data + cassette) in the front of the machine as well as a small keyboard. There were complaints with respect to the small keyboard, which soon led to the appearance of external replacement keyboards.

The PET 4000 series was launched in 1980, and the 4032 model was very successful at schools as its all-metal construction and all-in-one design made it ideal for the challenges in the classroom. The 4000 series used a larger 12" monitor and an enhanced BASIC 4.0 operating system. Commodore manufactured a successful variant called the *Teacher's* PET.

Fig. 11.4 Commodore PET 2001 home computer

Commodore introduced the 8000 series, and the last in the series was the SuperPET or SP9000. It used the Motorola 6809 microprocessor, and it provided support for several programming languages such as BASIC, Pascal, COBOL and FORTRAN. For more detailed information on Commodore, see [ORg:15].

11.6 Atari 400 and 800

Atari designed and produced four lines of home and personal computers from the late 1970s up to the early 1990s. These were the 8-bit Atari 400 and 800 line, the 16-bit ST line, the IBM PC compatible series and the 32-bit series.

The Atari 8-bit series began as a next-generation follow-up to its successful Atari 2600 video game console. Atari's management noted the success of Apple in the early personal computer market, and they tasked their engineers to transform the hardware into a personal computer system. The net result was the Atari 400 and the Atari 800 home computers, which were introduced in 1979.

The Atari 800 (Fig. 11.5) came with 8 KB of RAM and it retailed for $1000, and the Atari 400 was a lower-specification version, which retailed for $550. Both machines were based on the MOS 6502 microprocessor. The architecture of the Atari 400 and 800 machines provided sound and graphics capabilities that were superior to competitor products such as the Apple II or the Commodore PET.

The Atari 400 and 800 made an impact on the home computing field, and both machines included joystick ports for playing games. Atari BASIC was provided on an external cartridge for each machine.

The Atari 400 was Atari's entry-level computer, and it was designed for younger children. It had a membrane keyboard designed to prevent damage from food or small objects, and the keys could not be removed or swallowed by children. It was

Fig. 11.5 The Atari 800 home computer

initially designed for 4 K of memory, but as memory costs declined, it was shipped with 8 K (and later 16 K). This meant that it could run most cartridge- and cassette-based software. It was connected to a standard television.

The Atari 800 was based on the MOS 6502 microprocessor, and this 8-bit machine came with a graphics/audio chipset that allowed it to produce the most advanced graphics and sound on an existing home computer system. It could produce 128 colours (later upgraded to 256 colours using a later chip), and the graphics were 320 × 192, which was very advanced for its time. It looked like a standard typewriter machine.

The Atari 400 and 800 were replaced in 1982, initially with the Atari 1200XL and then with the Atari 600/800XL line of computers. For more detailed information on Atari, see [ORg:15].

11.7 Commodore 64

The Commodore 64 (C64) was a very successful 8-bit home computer introduced by Commodore in 1982 (Fig. 11.6). Its main competitors at that time were the Atari 400 and 800 and the Apple II computer. The cost of the C64 machine was $595, which was significantly less than its rivals, and Commodore cleverly exploited the price difference to rapidly gain market share. Approximately 15 million of the Commodore 64 machines were sold.

The C64 used the MOS 6501 microprocessor and it came with 64 kilobytes of RAM. It had 320 × 200 colour graphics with 16 colours using the VIC-II graphics chip, and the MOS Sound Interface Device (SID) chip. The SID chip was one of the first sound chips to be included in a home computer. The C64 dominated the low-end home computer market for most of the 1980s.

It came with the Commodore BASIC, but support for other languages such as Pascal and FORTRAN were also available. Programmers also wrote programs in assembly language to maximize speed and memory use. The Commodore 64's

Fig. 11.6 Commodore 64 home computer

graphics and sound capabilities were quite advanced for the time, and the C64 was very popular for computer games.

Commodore published detailed technical documentation to assist programmers and enthusiastic users to design and develop applications for the Commodore 64. This led to the development of over 10,000 commercial software applications such as development tools, games and office productivity applications for the machine. Atari was Commodore's main competitor, but it kept its technical information secret.

The C64 included a ROM-based version of the BASIC 2.0 programming language. There was no operating system as such, and instead the kernel was accessed via BASIC commands. BASIC did not allow commands for sound or graphics manipulation, and instead the user had to use the 'POKE' command to access these chips directly.

The Commodore 64 remained highly popular throughout the 1980s, and it was still being sold up to the early 1990s. For a more detailed account of Commodore, see [Bag:12].

11.8 Sinclair ZX 81 and ZX Spectrum

Sir Clive Sinclair founded Sinclair Research in 1973 as a consumer electronics company. It entered the home computer market in 1980 with the Sinclair ZX 80. This home computer retailed for £99.95, and it was the cheapest and smallest home computer in the United Kingdom, at the time.

The ZX 80 was a stepping stone for the Sinclair ZX 81 home computer, which was introduced in 1981. The ZX 81 was designed by Rick Dickinson to be a small, simple and low-cost home computer for the general public, and it retailed for an affordable £69.95. It offered tremendous value for money, and it opened the world of computing to those who had been denied access by cost. It was bought mainly for educational purposes.

The ZX 81 was a highly successful product with sales of over 1.5 million units. It came with 1 KB of memory, which could be extended to 16 KB of memory. It had a monochrome black-and-white display on a UHF television. It was one of the first home computers to be used widely by the general public, and it led to a large community of enthusiast users. It came with a BASIC interpreter, which enabled users to learn about computing and allowed them to write their first BASIC programs. It came with a standard QWERTY keyboard, which had some extra keys, and each key had several functions.

Sinclair entered an agreement with Timex, an American company, which allowed Timex to produce clones of Sinclair machines for the American market. These included the Timex Sinclair 1000 and the Timex Sinclair 1500 which were variants of the ZX 81. These were initially successful but soon faced intense completion from other American vendors.

The ZX Spectrum home computer (Fig. 11.7) was introduced in 1982, and it became the best-selling computer in the United Kingdom at that time. Its main competitor was the BBC Microcomputer produced by Acorn Computers. However, the BBC Micro was more expensive and retailed for £299, whereas the ZX Spectrum was about half its price. The basic model of the ZX Spectrum had 16 KB of RAM and retailed for £125, whereas the more advanced model had 48 KB of RAM and retailed for £175. This made the ZX Spectrum significantly more attractive to users than the existing Microcomputer.

The ZX Spectrum introduced colour graphics and sound, and it included an extended version of Sinclair's existing BASIC interpreter. It was an 8-bit home computer, and it used an 8-bit Zilog Z80 microprocessor. It initially came in two models and was eventually released in eight different models.

Rick Dickinson and Richard Altwasser designed the ZX Spectrum at Sinclair Research. Dickinson created the sleek outward design, and the internal hardware was designed by Altwasser. Clive Sinclair had emphasized the importance of creating a home computer substantially cheaper than the rival BBC Microcomputer, and so cost was a key factor in the design of the ZX Spectrum.

Cost forced the designers to find new ways of doing things, and they minimized the number of components in the keyboard from a few hundred to a handful of moving parts using a new technology. They used the cost-effective 3.5 MHz Z80 processor, a sound beeper, a BASIC interpreter and an audio tape as a storage device.

Fig. 11.7 ZX Spectrum

The demand for the ZX Spectrum was phenomenal, as the machine caught the imagination of the British general public. It was initially targeted as an educational tool to help students to become familiar with programming, but it soon became popular for playing home video games.

It was a highly successful home computer with over five million units sold. It was 50 % cheaper than the BBC Microcomputer and this was an important factor in its success. It led to a massive interest in learning about computing, programming and video games among the general public.

The users were supplied with a book from which they could type in a computer program into the computer, or they could load a program from a cassette. This allowed users to modify and experiment with programs as well as playing computer games.

Its simplicity, versatility and good design led to companies writing various software programs for it, and soon computer magazines and books dedicated to the ZX Spectrum were launched with the goal of teaching users how to program the machine.

The ZX Spectrum spawned various clones around the world. Countries such as the United States, to Russia and India created their own version of the Spectrum.

The ZX Spectrum remained popular throughout the 1980s, and it was officially retired in 1988. The Spectrum + was released in 1984, and this was essentially the 48 K version of the Spectrum with an enhanced keyboard. The Spectrum +128 was released in 1986, and it was similar in appearance to the Spectrum + but it had 128 K of memory.

Sinclair was sold to Amstrad in 1986, and Amstrad created its own models including the ZX Spectrum +2, the ZX Spectrum +2A, the ZX Spectrum +3, the ZX Spectrum +3A and the ZX Spectrum +3B.

There is a large archive of ZX Spectrum-related material available on-line (http://www.worldofspectrum.org), and it includes software, utilities, games and tools. Today, there are emulators available that allow Spectrum games to be downloaded and played on personal computers.

11.9 Apple Macintosh

The Apple Macintosh (Fig. 11.8) was announced during a famous television commercial aired during the third quarter of the Super Bowl in 1984. This was one of the most creative advertisements of all time, and it ran just once on television. It generated more excitement than any other advertisement up to then, and it immediately positioned Apple as a creative and innovative company, while implying that its competition (i.e. IBM) was stale and robotic.

It presented Orwell's totalitarian world of 1984, with a lady runner wearing orange shorts and a white tee shirt with a picture of the Apple Macintosh running towards a big screen and hurling a hammer at the big brother character on the screen. The audience is stunned at the broken screen and the voice-over states, 'On January 24th Apple will introduce the Apple Macintosh and you will see why 1984 will not

Fig. 11.8 Apple Macintosh computer (Photo public domain)

be like "1984".' Ridley Scott who has directed well-known films such as Alien, Blade Runner, Robin Hood and Gladiator directed the short film.

The Macintosh project began in Apple in 1979 with the goal of creating an easy-to-use low-cost computer for the average consumer. Jef Raskin initially led it, and the project team included Bill Atkinson, Burrell Smith and others. It was influenced by the design of the Apple Lisa, and it employed the Motorola 68000 processor. Steve Jobs became involved in the project in 1981 and Raskin left the project. Jobs negotiated a deal with Xerox that allowed him and other Apple employees to visit the Xerox PARC research centre at Palo Alto in California to see their pioneering work on the Xerox Alto computer and their work on a graphical user interface. PARC's research work had a major influence on the design and development of the Macintosh, as Jobs was convinced that future computers would use a graphical user interface. The design of the Macintosh included a friendly and intuitive graphical user interface (GUI), and the release of the Macintosh was a major milestone in computing.

The Macintosh was a much easier machine to use than the existing IBM PC. Its friendly and intuitive graphical user interface was a revolutionary change from the command-driven operating system of the IBM PC, which required the users to be familiar with its operating system commands. The introduction of the Mac GUI is an important milestone in the computing field, and it was 1990 before Microsoft introduced its Windows 3.0 GUI-driven operating system.

Apple intended that the Macintosh would be an inexpensive and user-friendly personal computer that would rival the IBM PC and compatibles. However, it was more expensive, and retailed for $2495, which was significantly more expensive that the IBM PC. Further, initially it had a limited number of applications available, whereas the IBM PC had spreadsheets, word processors and database applications, and so it was more attractive to customers. The technically superior Apple Macintosh was unable to break the IBM dominance of the market. However, the machine became very popular in the desktop publishing market, due to its advanced graphics capabilities.

11.10 Later Commodore and Atari Machines

The Amiga was a family of personal computers sold by Commodore in the 1980s and 1990s. Commodore purchased the start-up company called Amiga Corporation in 1984, and it became a subsidiary called Commodore-Amiga. The first model, the Amiga 1000 (or A1000), was released in 1985, and it became popular for its graphical, audio and multitasking capabilities. The A1000 had a powerful CPU and advanced graphics and sound hardware. It was based on the Motorola 68000 series of microprocessor, and it had 256 Kb of RAM, which could be upgraded with a further 256 Kb of RAM. It retailed for $1295.

The Amiga 500 (Fig. 11.9) was the best-selling model in the Amiga family, and it was released in 1987. It was a highly popular home computer with over 6 million machines sold. Several other models of the Amiga machines were introduced including the A3000; the A500+ and A600; and the A1200 and A4000 machines.

The August 1994 edition of the Byte magazine [By:94] spoke highly of the early Amiga machines. It called the A1000 machine the first multimedia computer, as it was so far ahead of its time with advanced graphics and sound.

Jack Tramiel (the founder and former CEO of Commodore) acquired Atari's home computing division in 1984, and he renamed the company to Atari Corporation. Atari designed the 16-bit GUI-based home computer, the Atari ST, in 1985. This machine was priced at an affordable $799, and it included a 360 KB

Fig. 11.9 Amiga 500 home computer (1987)

Fig. 11.10 Atari 1040 ST home computer

floppy disk drive, a mouse and a monochrome monitor. A colour monitor was provided for an extra $200, and the machine came with 512 KB of RAM. It used a colour graphical windowing system called GEM. The Atari ST included two Musical Instrument Digital Interface (MIDI) ports, which made it very popular with musicians.

The Atari 1040 ST (Fig. 11.10) was introduced in 1986, and this 16-bit machine differed from the Atari ST in that it integrated the external power supply and floppy disk drive into one case. It contained 1 MB of RAM and retailed for $999. It came as a complete system with a base unit, a monochrome monitor and a mouse. Atari released advanced versions of these models, called the Atari 520STE and the Atari 1040STE, in 1989. The Atari ST line had an impressive life span starting in 1986 and ending with the Atari Mega STE, which was released in 1990.

Atari released its first personal computer, the Atari PC, in 1987. This IBM-compatible machine was an 8 MHz 8088 machine with 512 KB of RAM and a 360 KB 5.25 inch floppy disk drive in a metal case. It released the Atari PC2 and PC3 later that year, and the PC3 included an internal hard disk. The Atari PC4 included a faster 16 MHz 80286 CPU and 1 MB of RAM, and it was released the same year. The PC5 was released in 1988 and it had a 20 MHz 80386 CPU and 2 MB of RAM.

The Atari ABC (Atari Business Computer) was released in 1990. The 286 version shipped with a range of CPU and storage choices ranging from an 8 MHz to a 20 MHz CPU and a 30 MB to 60 MB hard disk. The Atari ABC 386 version included a 20 MHz or 40 MHz CPU and a 40 MB or 80 MB hard disk. The ABC 386 shipped with Microsoft Windows 3.0. For more detailed information on Atari, see [Edw:11, IGN:14].

There was intense rivalry between the Amiga and Atari families of personal computers. However, ultimately both companies lost the battle in the personal computer market, and players such as IBM, Dell, HP and Apple now dominate it.

11.11 Review Questions

1. What is the significance of the Xerox Alto in the history of computing?
2. Discuss the relevance of Atari to game development and the computing field.
3. Discuss the accuracy of the message conveyed by Apple in the 1984 Super Bowl commercial that launched the Apple Macintosh.
4. Discuss whether Apple should have received all of the credit for its GUI-based operating system on the Macintosh given the pioneering work done at Xerox PARC?
5. Describe the relevance of the Apple I and II computers to the computing field.
6. Describe the relevance of the Sinclair Research to the computing field.
7. Explain the relevance of the MITS Altair 8800 to the computing field.

11.12 Summary

The invention of the microprocessor was a revolution in computing, and it led to the development of home and personal computers. Many of the early home computers discussed in this chapter were based on the 8-bit MOS 6502 microprocessor, with the MITS Altair 8800 based on the Intel 8080 microprocessor. Later home and personal computers used a variety of microprocessors such as the 8-bit Zilog Z80 microprocessor, the Motorola 68000 microprocessor, the Intel 8088 microprocessor and later Intel microprocessors.

We discussed the Xerox Alto computer, which was developed at Xerox PARC. This computer pioneered several key concepts in personal computing, and it had a major impact on the design of the Apple Macintosh. It was essentially a small minicomputer rather than a personal computer.

We discussed several early home computers such as the Apple I and II computers, which were developed by Apple; the Commodore PET which was introduced by Commodore Business Machines; the Atari 400 and 800 computers which were introduced by Atari; the Commodore 64 computer; the Apple Macintosh computer; the ZX Spectrum introduced by Sinclair Research; and later Atari and Amiga computers.

The IBM Personal Computer

12

Abstract

We discuss the introduction of the IBM personal computer, which was a major milestone in the computing field. The introduction of the IBM personal computer was a paradigm shift in that it placed computing power in the hands of millions of people. The previous paradigm was that an individual user had limited control over a computer, with the system administrators controlling the access privileges of the individual users. IBM's goal was to get into the home computer market as quickly as possible, and this led IBM to build the machine from off-the-shelf parts from a number of equipment manufacturers. IBM outsourced the development of the operating system to a small company called Microsoft, and Intel was chosen to supply the microprocessor for the IBM PC. Intel and Microsoft later became technology giants. The open architecture of the IBM PC led to a new industry of IBM-compatible computers.

Key Topics

Intel 8088
Intel 8086
PC/DOS
MS/DOS
IBM compatible
CP/M
Digital Research

G. O'Regan, *Introduction to the History of Computing*, Undergraduate Topics in Computer Science, DOI 10.1007/978-3-319-33138-6_12

12.1 Introduction

The introduction of the IBM personal computer in 1981 was a major milestone in the computing field. IBM's traditional approach up to then in product development was to develop a full proprietary solution. However, due to the aggressive timescales associated with the introduction of the IBM PC, it decided instead to outsource the development of the microprocessor to a small company called Intel and to outsource the development of the operating system to a small company called Microsoft. These decisions would later prove costly to IBM, as Microsoft and Intel later became technology giants.

The introduction of the IBM personal computer was a paradigm shift in computing in that it placed computing power in the hands of millions of people. The previous paradigm was that an individual user had limited control over a computer, with the system administrators controlling the access privileges of the individual users.

The awarding of the contract to develop the operating system to Microsoft later proved controversial. IBM had intended awarding the contract to Digital Research who had introduced the CP/M operating system for several microprocessors. However, IBM and Digital Research were unable to agree terms (there may have been problems with meeting the IBM delivery timescales or royalties demanded), and IBM instead awarded the contract to Microsoft. Microsoft hired a consultant to port an existing CP/M operating system to the 8088 microprocessor, and it later became clear to Digital Research that their software had been used to develop the operating system for the IBM personal computer.

12.2 The IBM Personal Computer

IBM introduced the IBM personal computer (PC) in 1981 as a machine to be used by small businesses and users in the home. The IBM goal at the time was to get quickly into the home computer market, which was then dominated by Commodore, Atari and Apple.

IBM assembled a small team of 12 people led by Don Estridge (Fig. 12.1), and their objective was to get the personal computer to the market as quickly as possible. They designed and developed the IBM PC within 1 year, and as time to market was the key driver, they built the machine with *off-the-shelf* parts from a number of equipment manufacturers. The normal IBM approach to the design and development of a computer was to develop a full proprietary solution

The team had intended using the IBM 801 processor, which was developed at the IBM Research Center in Yorktown Heights. However, they decided instead to use the Intel 8088 microprocessor, which was inferior to the IBM 801. They chose the PC/DOS operating system from Microsoft rather than developing their own operating system.

The unique IBM elements in the personal computer were limited to the system unit and keyboard. The team decided on an open architecture so that other manufacturers could produce and sell peripheral components and software without

Fig. 12.1 Don Estridge (Courtesy of IBM Archives)

Fig. 12.2 IBM personal computer (Courtesy of IBM Archives)

purchasing a licence. They published the *IBM PC Technical Reference Manual*, which included the complete circuit schematics, the IBM ROM BIOS source code and other engineering and programming information.

The IBM PC (Fig. 12.2) was the cheapest IBM computer produced up to then, and it was priced at an affordable $1565. It offered 16 kilobytes of memory (expandable to 256 kilobytes), a floppy disk, a keyboard and a monitor. The IBM personal computer became an immediate success, and it became the industry standard.

The open architecture led to a new industry of *IBM-compatible* computers, which had all of the essential features of the IBM PC, except that they were cheaper. The terms of the licensing of PC/DOS operating system gave Microsoft the rights to the MS/DOS operating system on the IBM-compatible computers, and this led inexorably to the rise of the Microsoft Corporation. The IBM Personal Computer XT was introduced in 1983. This model had more memory, a dual-sided diskette drive and a high-performance fixed-disk drive. The Personal Computer/AT was introduced in 1984.

The development of the IBM PC meant that computers were now affordable to ordinary users, and this led to a huge consumer market for personal computers and software. It led to the development of business software such as spreadsheets and accountancy packages, banking packages, programmer developer tools such as compilers for various programming languages, specialized editors and computer games.

The Apple Macintosh was announced in a famous television commercial aired during the Super Bowl in 1984. It was quite different from the IBM PC in that it included a friendly and intuitive graphical user interface, and the machine was much easier to use than the standard IBM PC. The latter was a command-driven operating system that required its users to be familiar with the PC/DOS commands. However, the Apple Macintosh was more expensive than the IBM PC, and cost proved to be a decisive factor for consumers when purchasing a personal computer. The IBM PC and the various IBM-compatible computers remained dominant.

The introduction of the personal computer was a paradigm shift in computing, and it led to a fundamental change in the way in which people worked. It placed computing power directly in the hands of millions of people, with individual users having complete control over the machine. The previous paradigm was that the system administrators strictly controlled the access privileges of the individual users, and so individual users had limited control over the computer. The introduction of the client-server architecture led to the linking of the personal computers (clients) to larger computers (servers). These servers contained large amounts of data that could be shared with the individual client computers.

The IBM strategy in developing the IBM personal computer was deeply flawed, and it cost the company dearly. IBM had traditionally produced all of the components for its machines, but with its open architecture model, any manufacturer could now produce an IBM-compatible machine. IBM had outsourced the development of the microprocessor chip to Intel, and Intel later became the dominant player in the microprocessor industry.

The development of the operating system, PC/DOS (PC Disk Operating System) was outsourced to a small company called Microsoft.[1] This proved to be a major mistake by IBM, as the terms of the deal with Microsoft were favourable to the lat-

---

[1] Microsoft was founded by Bill Gates and Paul Allen in 1975.

ter, and it allowed Microsoft to sell its own version of the operating system (i.e. MS/DOS) to other manufacturers as the operating system for the many IBM compatibles. Intel and Microsoft would later become technology giants.

12.3 Operating System for IBM PC

Digital Research lost out on the opportunity of a lifetime to supply the operating for the IBM personal computer to IBM, and instead it was Microsoft that reaped the benefits. Microsoft would later become a technology giant and a dominant force in the computer industry.

Bloomberg Businessweek published an article in 2004 describing the background to the development of the operating system for the IBM PC and the failed negotiations between Digital Research and IBM on the licensing of the CP/M operating system. The article was titled *The man who could have been Bill Gates* [Blo:04].

Don Estridge led the IBM team that was developing the IBM personal computer. The project was subject to an aggressive delivery schedule, and while traditionally IBM developed a full proprietary solution, it decided instead to outsource the development of the microprocessor and the operating system.

The IBM team initially asked Bill Gates and Microsoft in Seattle to supply them with an operating system. Microsoft had already signed a contract with IBM to supply a BASIC interpreter for the IBM PC, but they lacked the appropriate expertise in operating system development. Gates referred IBM to Gary Kildall at DRI, and the IBM team approached Digital Research with a view to licensing its CP/M operating system.

Digital Research was working on a new version of CP/M for the 16-bit Intel 8086 microprocessor, which had been introduced in 1978. IBM decided to use the lower-cost Intel 8088 microprocessor (a slower version of the 8086) for its new personal computer.

IBM and Digital Research failed to reach an agreement on the licensing of CP/M for the IBM PC. The precise reasons for failure are unclear, but some immediate problems arose with respect to the signing of an IBM non-disclosure agreement during the visit. It is unclear whether Kildall actually met with IBM and whether there was an informal handshake agreement between both parties. However, there was certainly no documented legal agreement between IBM and DRI.

There may also have been difficulties in relation to the amount of royalty payment being demanded by Digital Research, as well as practical difficulties in achieving the required IBM delivery schedule (due to Digital Research's existing commitments to Intel). Kildall was superb at technical innovation, but he may have lacked the appropriate business acumen to secure a good deal or he may have oversold his hand.

Gates had signed a Microsoft BASIC licence agreement with IBM, and he now saw a business opportunity for Microsoft. He offered to provide an operating system (later called PC/DOS) and BASIC to IBM on favourable terms. IBM accepted the offer, and the contract allowed Microsoft to market and sell its version (MS/DOS) of the operating systems on IBM compatibles. Microsoft reaped the benefits.

Gates was aware of the work done by Tim Paterson on a simple quick-and-dirty version of CP/M (called QDOS) for the 8086 microprocessor for Seattle Computer Products (SCP). Gates licensed QDOS for $50,000, and he hired Paterson to modify it to run on the IBM PC for the Intel 8088 microprocessor. Gates then licensed the operating system to IBM for a low per-copy royalty fee.

IBM called the new operating system PC/DOS, and Microsoft retained the rights to MS/DOS, which was used on IBM-compatible computers produced by other hardware manufacturers. In time, MS/DOS would later become the dominant operating system (eclipsing PC/DOS due to the open architecture of the IBM PC and the rapid growth of clones) leading to the growth of Microsoft into a major corporation.

DRI released CP/M-86 shortly after IBM released PC DOS. Kildall examined PC/DOS, and it was clear to him that it had been derived from CP/M. He was furious and met separately with IBM and Microsoft, but nothing was resolved. Digital Research considered suing Microsoft for copying all of the CP/M system calls in DOS 1.0, as it was evident to Kildall that Paterson's QDOS was a copy of CP/M.

He considered his legal options, but his legal advice suggested that as intellectual copyright law with regard to software had only been recently introduced in the United States, it was not clear what constituted infringement of copyright. There was no guarantee of success in any legal action against IBM, and considerable expense would be involved. Kildall threatened IBM with legal action, and IBM agreed to offer both CP/M-86 and PC-DOS. However, as CP/M was priced at $240 and DOS at $60, few personal computer owners were willing to pay the extra cost. CP/M was to fade into obscurity.

Perhaps, if Kildall had played his hand differently, he could have been in the position that Bill Gates is in today, and Digital Research could well have been the *Microsoft* of the PC industry. Kildall's delay in developing the operating system gave Paterson the opportunity to create his own version. IBM was under serious time pressures with the development of the IBM PC, and Kildall may have been unable to meet the IBM deadline. This may have resulted in IBM dealing with Gates instead of DRI.

Further, the size of the royalty fee demanded by Kildall for CP/M was not very sensible, as the excessive fee resulted in very low sales for the DRI product, whereas if a more realistic price had been proposed, then DRI may have made some reasonable revenue. Nevertheless, Kildall could justly feel hard done by, and he may have viewed Microsoft's actions as the theft of his intellectual ideas and technical inventions. It shows that technical excellence and innovation are not in itself sufficient for business success.

12.4 Review Questions

1. Why did IBM launch the personal computer ?
2. What mistakes did IBM make with its introduction of the IBM PC?
3. Why has Gary Kildall been described as "the man who could have been Bill Gates"?
4. Describe the controversy over the operating system for the IBM PC.
5. Describe IBM's contributions to the computing field.
6. Describe Intel's contributions to the computing field.
7. Describe Microsoft's contributions to the computing field.

12.5 Summary

The introduction of the IBM personal computer in 1981 was a major milestone in the computing field. IBM's approach up to then was to develop a full proprietary solution. However, due to the timescales associated with the development of the IBM PC, it decided instead to outsource the development of the microprocessor to a small company called Intel and to outsource the development of the operating system to a small company called Microsoft.

Don Estridge led the IBM team responsible for the introduction of the IBM PC, and their goal was to design and develop the IBM PC within 1 year. They built the machine with *off-the-shelf* parts from a number of equipment manufacturers, rather than the usual IBM approach developing a full proprietary solution.

The awarding of the contract to develop the operating system to Microsoft later proved controversial. IBM had intended awarding the contract to Digital Research who had introduced the CP/M operating system for several microprocessors.

A Short History of Telecommunications 13

Abstract

Telecommunications is a branch of technology concerned with the transmission of information over a distance, where the transmitter sends the information to a receiver. We present a short history of telecommunications and focus on the development of mobile phone technology. The development of the AXE system by Ericsson is discussed, and this was the first fully automated digital switching system. We discuss the concept of a cellular system, which was introduced by Bell Labs, as well as the introduction of the first mobile phone, the DynaTAC, by Motorola. We discuss the Iridium system, which was launched in late 1998 to provide worldwide wireless coverage to its customers, and the coverage included the oceans, airways and polar regions. The existing telecom systems had limited coverage in remote areas, and so the concept of global coverage as provided by Iridium was potentially very useful. In many ways, Iridium was an engineering triumph over common sense, and over $5 billion was spent in building an infrastructure of low Earth orbit (LEO) satellites to provide global coverage.

Key Topics

Telegraph
Telephone
AMPS
AXE
Telephone
Telegraph
Mobile phone system
Iridium

G. O'Regan, *Introduction to the History of Computing*, Undergraduate Topics in Computer Science, DOI 10.1007/978-3-319-33138-6_13

13.1 Introduction

Telecommunications is a branch of technology concerned with the transmission of information over a distance, where the transmitter sends the information to a receiver. Early societies used fire and smoke signals for visual communication, with drums used for auditory communication. This allowed simple messages (e.g. 'danger') to be communicated to other groups.

The Persian Empire established an early postal system in the sixth century B.C., and other societies such as the Egyptians and Romans later established their own postal systems. A pigeon messaging system, where the homing characteristics of pigeons were employed to send messages, was later introduced.

The Greeks introduced an early semaphore system in the fourth century B.C., and this allowed very simple messages to be exchanged between groups on two different hills (similar in a sense to smoke signals). A ship semaphore system was introduced in the fifteenth century, which allowed two ships to communicate with each other. This system used flags where the position and motion of a flag represented a letter.

The Chappe brothers in France introduced an early optical telegraph system in Europe in the late eighteenth century. It used similar principles as the ship-based semaphore system, and it allowed messages to be sent from one high tower to another. It was used by the French military.

Early electrical telegraph systems were introduced in the early nineteenth century, and Samuel Morse devised a system (the Morse code) that allowed letters to be represented by a series of on-off tones in the late 1830s. This was the foundation for electrical telegraphs and later telephone systems. The first Atlantic telegraph cable was laid between Britain and America in 1858.

The telephone was invented by Alexander Graham Bell in 1876,[1] and early telephones were hardwired to and communicated with a single other telephone (e.g. from a person's business to his home), as initially there were no telephone exchanges. A telephone exchange provides switching or interconnection between two subscriber lines, and the earliest manual commercial telephone exchanges were introduced in the late 1870s. The first mechanical automated exchanges were introduced in the early 1890s. The first North American transcontinental phone call from the east coast to the west coast was made by Bell in 1915, and it made long-distance communication a reality.

The invention of the telephone was a paradigm shift from *face-to-face* communication, where people met to exchange ideas and share information or where individuals wrote letters to each other to exchange information. The telephone was a new medium that provided direct and instantaneous communication between two people. It allowed two individuals to establish and maintain two-way communication irrespective of being at two different physical locations. Initially the business

[1] He was the first person to patent the telephone as an 'apparatus for transmitting vocal or other sounds telegraphically'. There are several other claimants for inventing the telephone.

community and the affluent members of society used the telephone, but this changed rapidly in the years that followed.

Marconi, an Italian engineer, introduced a system for the wireless transmission of sounds in 1896, and the British Marconi Company was established in 1897. It began communication between ships at sea and coastal radio stations. Marconi established an early radio factory in England in 1912.

The first prototype electronic television was developed and demonstrated by Philip Farnsworth in the late 1920s. It was the result of research on ways to transmit images, and it had been determined that radio waves could be encoded with an image and then transmitted back to the screen. Farnsworth's prototype is considered the first electronic television.

The foundations of the mobile cellular industry go back to the introduction of a limited-capacity mobile phone system that was introduced for automobiles in 1946. Martin Cooper of Motorola made the first mobile phone call to Joe Engels at Bell Labs in 1973, and a prototype mobile phone network was operational in the late 1970s with commercial mobile phone networks introduced in the early 1980s. The first global mobile phone system (Iridium) was operational in 1998, and the Iridium system consisted of 66 satellites, with the customers using hand-held satellite phones.

The ARPANET packet switching network was introduced in the late 1960s, and it remained operational until 1990, when the Internet became operational. The Internet has led to almost instantaneous communication, and it has led to electronic mail; the World Wide Web, which was developed by Tim Berners-Lee at CERN; social networking; electronic commerce; and telephone calls over the Internet with the VoIP protocol.

This chapter considers a small number of events in the history of telecommunications including the development of the AXE system, which was the first fully automated digital switching system, the development of mobile phone technology and the development of the Iridium satellite mobile phone system.

13.2 AXE System

Ericsson introduced the AXE (Automatic Exchange Electric) switching system in 1977 (Fig. 13.1). This was the first fully automated digital switching system, and it converted speech into digital (i.e., the binary language used by computers). Ericsson's competitors were still using the slower and less reliable analog systems.

The analog system uses an electric current to convey the vibrations of the human voice, whereas a digital system uses a stream of binary digits to represent sound. The AXE system was an immediate success with telecom companies, and it has been sold in many countries around the world. AXE was originally a digital exchange for landline telephony, but it has been extended for use with mobile telephony systems.

Ellemtel was established in 1970 as a pure research and development company and was a joint venture between Televerket (Sweden's state-owned PTT) and

Fig. 13.1 AXE system (Courtesy of Ericsson)

Ericsson. Its primary task was to develop an electronic and automated switching system for telephone stations that would become the AXE system.

Ericsson had been working to develop a commercial electronic switching system called AKE, while Televerket was working on its own electronic switch. Ericsson realized that its AKE system was not suitable for large switching stations and that it needed to develop a new generation of switching systems. It decided to combine its resources with Televerket and to jointly develop an electronic telephone switching system.

Bengt-Gunnar Magnusson was the project manager for the AXE project, and AXE had a modular system design which made the system flexible. New functionality could be added and existing modules updated or replaced. The modular design enabled the system to be easily adapted to different markets.

The development of AXE also involved the development of hardware and software such as programs and processors to control the AXE stations. The first prototype AXE system was installed at a Televerket station in 1976, and Ellemtel's work in developing the AXE system was complete in 1978.

The AXE system was then commercialized and many of Ellemtel's employees moved to Ericsson. AXE was an immediate success and Ericsson soon had customers in Sweden, Finland, France, Australia and Saudi Arabia. The Saudi order was the largest that Ericsson had ever received, and it involved increasing the capacity of the Saudi network by 200 % and installing the AXE system.

The introduction of AXE meant that by the early 1980s, Ericsson had the market's most advanced and flexible switching system, and this made it ideally placed for the transition to mobile telephony. It meant that Ericsson had moved from being a minor player in the telecoms business to a major league player. It was now the leader in fixed-line phone technology, and it laid the foundation for Ericsson's future success in mobile telephony, where it became the leader in mobile technology from the late 1980s. Today, AXE has been installed in over 130 countries.

13.3 Development of Mobile Phone Standards

Bell Labs played an important role (with Motorola) in the development of the analog mobile phone system in the United States. It developed a system in the mid-1940s that allowed mobile users to place and receive calls from automobiles, and Motorola developed mobile phones for automobiles. However, these phones were large and bulky and they consumed a lot of power. A user needed to keep the automobile's engine running in order to make or receive a call.

Bell Labs first proposed the idea of a cellular system back in the late 1940s, when they proposed hexagonal rings for mobile communication. Large geographical areas were divided into cells, where each cell had its own base station and channels. The available frequencies could be used in parallel in different cells without disturbing each other (Fig. 13.2). Mobile telephone could now, in theory, handle a large number of subscribers. However, it was not until the late 1960s that Bell Labs prepared a detailed plan for implementing the cellular system.

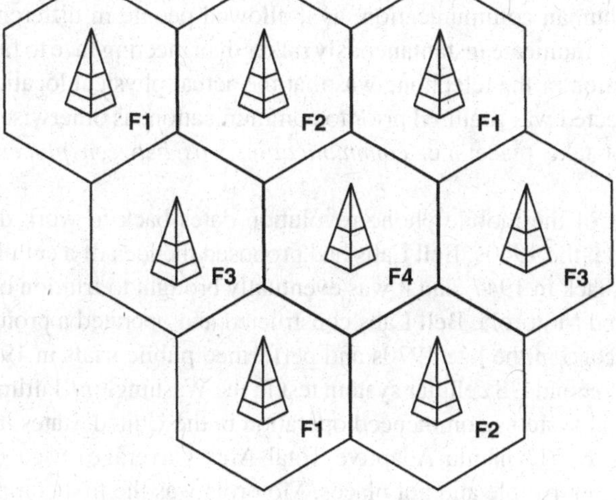

Fig. 13.2 Frequency reuse in cellular networks

Bell Labs developed the Advanced Mobile Phone System (AMPS) standard from 1968 to 1983. Motorola and other telecommunication companies designed and built phones for this cellular system. AMPS uses separate frequencies (or channels) for each conversation and requires considerable bandwidth for a large number of users.

The signals from a transmitter cover an area called a cell. As a user moves from one cell into a new cell, a handover to the new cell takes place without any noticeable difference to the user. The signals in the adjacent cell are sent and received on different channels to the existing cell's signals, and so there is no interference.

The Total Access Communication System (TACS) and Extended TACS (ETACS) were variants of AMPS that were employed in the United Kingdom and Europe. These analog standards employed separate frequencies (or channels) for each conversation using frequency division multiple access (FDMA). However, the analog system suffered from static and noise, and there was no protection from eavesdropping using a scanner.

Ericsson became the leader in the first generation of mobile with Motorola, and the extent of its leadership was clear when its proposed design for digital mobile radio transmission was selected as the US standard for cellular communications over entries from Motorola and AT&T in 1989.

AMPS is the first generation of cellular technology, and it has several weaknesses when compared to today's cellular systems. Mobile technology has evolved from the AMPS analog standard to the digital Global System for Mobile communication (GSM) and code division multiple access (CDMA) technologies; to General Packet Radio Service (GPRS); to third-generation mobile, including 3G and WCDMA; and to fourth-generation (4G) mobile.

13.4 Development of Mobile Phone Technology

The invention of the telephone by Graham Bell in the late nineteenth century was a revolution in human communication, as it allowed people in different geographic locations to communicate instantaneously rather than meeting face to face. However, the key restriction of the telephone was that the actual physical location of the person to be contacted was required prior to communication, as otherwise communication could not take place: i.e. *communication was between places rather than people*.

The origins of the mobile phone revolution dates back to work done on radio technology from the 1940s. Bell Labs had proposed the idea of a cellular communication system back in 1947, and it was eventually brought to fruition by researchers at Bell Labs and Motorola. Bell Labs constructed and operated a prototype cellular system in Chicago in the late 1970s and performed public trials in 1979. Motorola commenced a second US cellular system test in the Washington/Baltimore area. The first commercial systems commenced operation in the United States in 1983.

The DynaTAC (Dynamic Adaptive Total Area Coverage) used cellular radio technology to link people and not places. Motorola was the first company to incorporate the technology into a portable device designed for use outside of an automobile, and it spent $100 million on the development of cellular technology. Martin

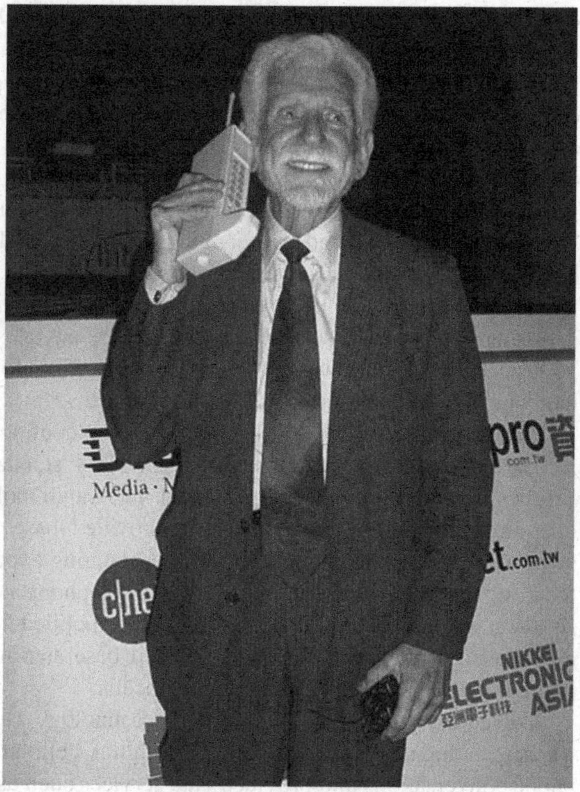

Fig. 13.3 Martin Cooper re-enacts DynaTAC call

Cooper (Fig. 13.3) led the team at Motorola that developed the DynaTAC 8000X, and he made the first mobile phone call on a prototype DynaTAC phone to Joel Engels, the head of research at Bell Labs, in April 1973.

Commercial cellular services commenced in North America in 1983, and the world's first commercial mobile phone went on sale the same year. This was the Motorola DynaTAC 8000X, and it was popularly known as the *brick* due to its size and shape. It weighed 28 ounces (almost 2 lbs); it was 13.5" (over a foot) in length and 3.5" in width. It had a LED display and could store 30 numbers. It had a talk time of 30 min and 8 h of standby, and it took over 10 h to recharge.

The cost of the Motorola DynaTAC 8000X was $3995, and it was too expensive for most people apart from wealthy consumers. Today, mobile phones are ubiquitous, and there are more mobile phone users than fixed-line users. The cost of a mobile phone today is typically less than $100, and a mobile phone typically weighs as little as 3 ounces.

The first-generation mobile phone system introduced into North America in the early 1980s used the 800 MHz cellular band. It had a frequency range between 800 and 900 MHz. Each service provider could use half of the 824–849 MHz range for receiving signals from cellular phones and half the 869–894 MHz range for transmitting to cellular phones. The bands were divided into 30 kHz sub-bands called channels, and a separate frequency (or channel) was used for each conversation. The division of the spectrum into sub-band channels is achieved by using frequency division multiple access (FDMA).

This first-generation system allowed voice communication only, and it was susceptible to static and noise. Further, it had no protection from eavesdropping using a scanner.

The AXE system (discussed earlier) provided the foundation for Ericsson's growth in mobile telephony. The flexible modular design of AXE allowed new functionality to be added, and by changing a module, AXE could be reconfigured to handle mobile telephone calls. This allowed Ericsson to design the first mobile telephone exchange (MTX) by replacing the subsystem for fixed subscribers with a new subsystem for mobile subscribers. The MTX switch was developed in the late 1970s/early 1980s and was a key part of the Nordic Mobile Telephone (NMT) system which would be used in all Nordic countries.

Ericsson was awarded a large Saudi Arabian contract to deliver a fixed-line and mobile system, and it was agreed that the NMT standard would be used and that Ericsson would supply the entire system. The Saudi mobile phone network became operational from 1981, and Ericsson provided base stations, radio towers and switches. Ericsson had now acquired cell-planning experience, and it was awarded the contract to develop the entire mobile telephone network in the Netherlands. Ericsson was now a total systems supplier in mobile telephony, and it provided the entire infrastructure such as switches and base stations. Today, its base stations range from small picocells to large macrocells.

The second generation (2G) of mobile technology was a significant improvement on the existing analog technology. This digital, cellular technology encrypted telephone conversations and provided data services such as text and picture messages. The two main second-generation technologies were the GSM standard developed by the European Telecommunications Standards Institute (ETSI) and CDMA developed in the United States. The first GSM call was made by the Finnish prime minister in Finland in 1991, and the first short message service (SMS) or text message was sent in 1992.

The subscriber identity module (SIM) card was a new feature in GSM, and a SIM card is a detachable smart card that contains the user's subscription information and phone book. The SIM card may be used in other GSM phones, and this is useful when the user purchases a replacement phone. GSM provides an increased level of security, with communication between the subscriber and base station encrypted.

GSM networks evolved into GPRS (2.5 G), which became available in 2000. Third- and fourth-generation (3G and 4G) mobiles provide mobile broadband multimedia communication. Mobile phone technology has transformed the earlier paradigm of *communication between places* to that of *communication between people*.

Motorola dominated the analog mobile phone market. However, it was slow to adapt to the GSM standard, and it paid a heavy price with a loss of market share to

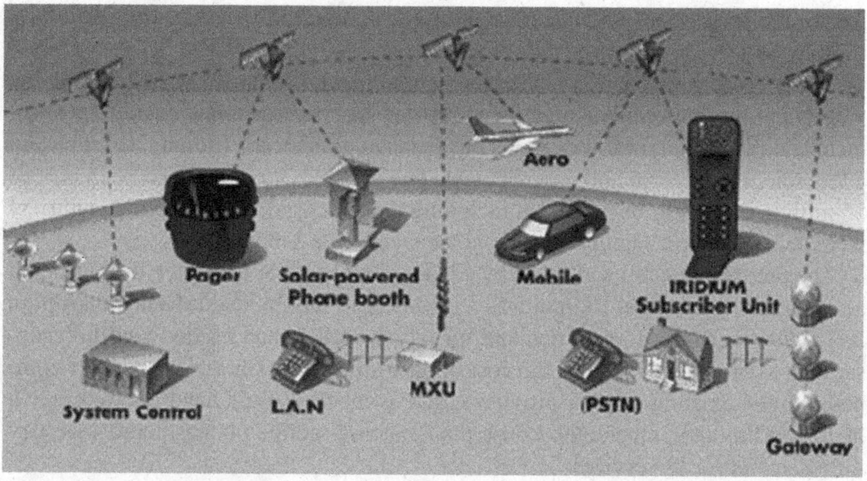

Fig. 13.4 Iridium system (Courtesy of Iridium)

Nokia and Ericsson. The company was very slow to see the potential of a mobile phone as a fashion device,[2] and it was too slow in adapting to smartphones.

13.5 The Iridium Satellite System

Iridium was a global satellite phone company that was backed by Motorola. In many ways it was an engineering triumph over common sense, and over $5 billion was spent in building an infrastructure of low Earth orbit (LEO) satellites to provide global coverage. It was launched in the late 1998 to provide worldwide wireless coverage to its customers, and the coverage included the oceans, airways and polar regions. The existing telecom systems had limited coverage in remote areas, and so the concept of global coverage as provided by Iridium was potentially very useful.

Iridium was implemented by a constellation of 66 satellites (Fig. 13.4). The original design required 77 satellites, and so the name *Iridium* was chosen since its atomic number in the periodic table is 77. However, the later design required just 66 satellites, and so *Dysprosium* may have been a more appropriate name. The satellites are in low Earth orbit at a height of approximately 485 miles, and communication between the satellites is via inter-satellite links. Each satellite contains seven Motorola Power PC 603E processors running at 200 MHz. These machines are used for satellite communication and control.

Iridium routes phone calls through space and there are several Earth stations. As satellites leave the area of an Earth base station, the routing tables change, and

[2] The attitude of Motorola at the time seemed to be similar to that of Henry Ford: i.e. they can have whatever colour they like as long as it is black.

frames are forwarded to the next satellite just coming into view of the Earth base station.

The Iridium constellation is the largest commercial satellite constellation in the world, and it is especially suited for industries such as maritime, aviation, government and the military. Motorola was the prime contractor for Iridium, and it played a key role in its design and development. The satellites were produced at a cost of $5 million each ($40 million each including launch costs), and Motorola engineers were able to make a satellite in a phenomenal time of 2–3 weeks.

The first Iridium call was made by Al Gore in late 1998. However, despite being an engineering triumph, Iridium was a commercial failure, and it went bankrupt in late 1999 due to insufficient demand for its services. It had needed a million subscribers to break even, and as the cost of an Iridium call was very expensive compared to the existing cellular providers, and as the cost of its handsets were much higher and more cumbersome to use than existing mobile phones, there was very little demand for its services.

Specifically, the reasons for failure included:

- Insufficient demand for its services (10,000 subscribers)
- High cost of its service ($5 per minute for a call)
- Cost of its mobile handsets ($3000 per handset)
- Bulky mobile handsets
- Competition from existing mobile phone networks
- Management failures

However, the Iridium satellites remained in orbit, and the service was re-established in 2001 by the newly founded Iridium Satellite LLC. The new business model required just 60,000 subscribers to break even. Today, it has over half a million customers, and it is used extensively by the US Department of Defense.

Iridium was designed in the late 1980s, and so it is designed primarily for voice rather than data. This means that it lacks the sophistication of modern mobile phone networks, and it is not as attractive to users. However, it provides service in remote parts of the world, which is very useful.

13.6 Review Questions

1. Describe the contributions of Bell Labs to mobile technology.
2. What are the advantages of mobile technology over fixed-line technology?
3. Describe the various generations of mobile technology.
4. Describe Motorola's contributions to mobile technology.
5. What factors led to Ericsson's success and leadership in mobile technology?
6. What factors led to the (initial) commercial failure of the Iridium system?

13.7 Summary

The invention of the telephone by Graham Bell in the late nineteenth century was a revolution in human communication, as it allowed people in different geographic locations to communicate instantaneously rather than meeting face to face. The early phones had major limitations, but the development of automated telephone exchanges helped to deal with many of these.

However, the key restriction of the telephone was that the actual physical location of the person to be contacted was needed prior to communication: i.e. communication was between places rather than people.

This led to research by Bell Labs and others into ways in which communication could take place between people (and not places). Bell Labs developed a system in the mid-1940s that allowed mobile users to place and receive calls from automobiles, with Motorola developing the phones for automobiles. However, these phones were large and bulky, and the automobile's engine needed to be running in order to make or receive a call.

Bell Labs proposed the idea of a cellular system back in the late 1940s, and it prepared a detailed plan for the cellular system in the late 1960s. A cellular system is divided into cells, where each cell has its own base station and channels. The available frequencies may be used in parallel in different cells without interference with each other.

Motorola developed the first mobile phone, the DynaTAC, and it made the first mobile phone call in 1973. The first mobile phone systems were analog and based on the AMPS standard. The later generations of mobile technology are digital and are a significant advance on the older cellular technology.

Iridium provides global wireless coverage to its customers including coverage in the oceans, airways and polar regions. It was implemented by a constellation of 66 satellites. For a more detailed account of the contributions of Bell Labs, Ericsson and Motorola, see [Ger:13, MeJ:01, Mot:99, ORe:15].

The Internet Revolution

14

Abstract

This chapter describes the Internet revolution starting from ARPANET, which was a packet-switched network, to TCP/IP, which is a set of network standards for interconnecting networks and computers. These developments led to the birth of the Internet, and Tim Berners-Lee's work at CERN led to the birth of the World Wide Web. Berners-Lee built on several existing inventions such as the Internet, hypertext and the mouse to form the World Wide Web. Applications of the World Wide Web are discussed, as well as successful and unsuccessful new economy companies. The dot-com bubble and subsequent burst of the late 1990s/early 2000 are discussed.

Key Topics

ARPANET
TCP/IP
The Internet
The World Wide Web
Dot-com bubble
E-software development
Facebook
The Twitter Revolution

14.1 Introduction

The vision of the Internet and World Wide Web goes back to an article by Vannevar Bush in the 1940s. Bush was an American scientist who had done work on submarine detection for the US Navy. He designed and developed the differential analyser

G. O'Regan, *Introduction to the History of Computing*, Undergraduate Topics in Computer Science, DOI 10.1007/978-3-319-33138-6_14

Fig. 14.1 Vannevar Bush

(Fig. 1.1), which was a mechanical computer whose function was to evaluate and solve first-order differential equations. It was funded by the Rockefeller Foundation and developed by Bush and others at MIT in the early 1930s. Bush supervised Claude Shannon at MIT, and Shannon's initial work was to improve the differential analyser.

Bush (Fig. 14.1) became the director of the office of Scientific Research and Development, and he developed a win-win relationship between the US military and universities. He arranged large research funding for the universities to carry out applied research to assist the US military. This allowed the military to benefit from the early exploitation of research results, and it also led to better facilities and laboratories at the universities. It led to close links and cooperation between universities such as Harvard and Berkeley, and this would eventually lead to the development of ARPANET by DARPA.

Bush outlined his vision of an information management system called the *memex* (memory extender) in a famous essay *As We May Think* [Bus:45]. He envisaged the memex as a device electronically linked to a library that would be able to display books and films. It describes a proto-hypertext computer system and influenced the later development of hypertext systems.

> A memex is a device in which an individual stores all his books, records, and communications, and which is mechanized so that it may be consulted with exceeding speed and flexibility. It is an enlarged intimate supplement to his memory.

> It consists of a desk, and while it can presumably be operated from a distance, it is primarily the piece of furniture at which he works. On the top are slanting translucent screens, on which material can be projected for convenient reading. There is a keyboard, and sets of buttons and levers. Otherwise it looks like an ordinary desk.

Bush predicted that:

> Wholly new forms of encyclopedias will appear, ready made with a mesh of associative trails running through them, ready to be dropped into the memex and there amplified.

This description motivated Ted Nelson and Douglas Engelbart to independently formulate ideas that would become hypertext. Tim Berners-Lee would later use hypertext as part of the development of the World Wide Web.

14.2 The ARPANET

There were approximately 10,000 computers in the world in the 1960s. These were expensive machines (over $100 K) with limited processing power. They contained only a few thousand words of magnetic memory, and programming and debugging was difficult. Further, communication between computers was virtually non-existent.

However, several computer scientists had dreams of worldwide networks of computers, where every computer around the globe is interconnected to all of the other computers in the world. Licklider[1] wrote memos in the early 1960s on his concept of an intergalactic network. This concept envisaged that everyone around the globe would be interconnected and able to access programs and data at any site from anywhere.

The US Department of Defense founded the Advanced Research Projects Agency (ARPA) in the late 1950s. ARPA embraced high-risk, high-return research, and Licklider became the head of its computer research program. He developed close links with MIT, UCLA and BBN Technologies.[2] The concept of packet switching[3] was invented in the 1960s, and several organizations including the National Physical Laboratory (NPL), RAND Corporation and MIT commenced work on its implementation.

The early computers had different standards for data representation, and so it was essential to know the standard employed by each computer prior to communication.

---

[1] Licklider was an early pioneer of AI and wrote an influential paper 'Man-Computer Symbiosis' in 1960, which outlined the need for simple interaction between users and computers.

[2] BBN Technologies (originally Bolt Beranek and Newman) is a research and development technology company. It played an important role in the development of packet switching and in the implementation and operation of ARPANET. The '@' sign used in an email address was a BBN innovation.

[3] Packet switching is a message communication system between computers. Long messages are split into packets, which are then sent separately so as to minimize the risk of congestion.

This led to recognition of the need for common standards in data representation, and a US government committee developed the American Standard Code for Information Interchange (ASCII) in 1963. This was the first universal standard for data, and it allowed machines from different manufacturers to exchange data. The standard allowed a 7-bit binary number to stand for a letter in the English alphabet, an Arabic numeral or a punctuation symbol. The use of 7 bits allowed 128 distinct characters to be represented. The development of the IBM System/360 mainframe (discussed in Chap. 8) standardized the use of 8 bits for a word, and 12-bit or 36-bit words became obsolete.

The first wide-area network connection was created in 1965,[4] and it involved the connection of a computer at MIT to a computer in Santa Monica. This was done via a dedicated telephone line, and it showed that a telephone line could be used for data transfer. ARPA recognized the need to build a network of computers in the mid-1960s, and this led to the ARPANET project in 1966 which aimed to implement a packet-switched network with a network speed of 56 Kbps. ARPANET was to become the world's first packet-switched network.

BBN Technologies was awarded the contract to implement the network, with plans for a total of 19 nodes. The first two nodes were based at UCLA and the Stanford Research Institute (SRI). The network management was performed by interconnected *Interface Message Processors* (IMPs), which were in front of the main computers. The IMPs eventually evolved to become the network routers that are used today

The team at UCLA called itself the *Network Working Group*, and it saw its role as developing a set of rules that specified how the computers on the network should communicate. These rules were called the *Network Control Protocol* (NCP). The first host-to-host connection was made between a computer in UCLA and a computer at SRI in late 1969. Several other nodes were added to the network until it reached its target of 19 nodes in 1971.

The Network Working Group developed the *telnet protocol* and the *File Transfer Protocol* (FTP) in 1971. The telnet program allowed the user of one computer to remotely log in to the computer of another computer. The File Transfer Protocol allows the user of one computer to send (or receive) files to (or from) another computer. A public demonstration of ARPANET was made in 1972 and it was a huge success. One of the earliest demos was that of Weizenbaum's ELIZA program (discussed in Chap. 19). This famous AI program allowed a user to conduct a typed conversation with an artificially intelligent machine (psychiatrist) at MIT.

The viability of packet switching as a standard for network communication had been clearly demonstrated. Ray Tomlinson of BBN Technologies developed a program that allowed electronic mail to be sent over the ARPANET. Over 30 institutions were connected to the ARPANET by the early 1970s.

[4] We will not consider the early work done by SAGE in the late 1950s.

14.3 TCP/IP

ARPA was renamed to the Defense Advanced Research Projects Agency (DARPA) in 1973. It commenced a project to connect seven computers on four islands using a radio-based network and a project to establish a satellite connection between a site in Norway and in the United Kingdom. This led to a need for the interconnection of the ARPANET with other networks. The key problems were to investigate ways of achieving convergence between ARPANET, radio-based networks and the satellite networks, as these all had different interfaces, packet sizes and transmission rates. Therefore, there was a need for a *network-to-network connection protocol*.

An international network-working group (INWG) was formed in 1973. The concept of the transmission control protocol (TCP) was developed at DARPA by Bob Kahn and Vint Cerf, and they presented their ideas at an INWG meeting at the University of Sussex in England in 1974 [KaC:74]. TCP allowed cross network connections, and it began to replace the original NCP protocol that was used in ARPANET.

TCP is a set of network standards that specify the details of how computers communicate, as well as the standards for interconnecting networks and computers. It was designed to be flexible and provides a transmission standard that deals with physical differences in host computers, routers and networks. It is designed to transfer data over networks which support different packet sizes and which may sometimes lose packets. It allows the internetworking of very different networks, which then act as one network.

The new protocol standards were known as the *Transport Control Protocol* (TCP) and the *Internet Protocol* (IP). TCP details how information is broken into packets and reassembled on delivery, whereas IP is focused on sending the packet across the network. These standards allow users to send electronic mail or to transfer files electronically, without needing to concern themselves with the physical differences in the networks. TCP/IP consists of four layers (Table 14.1).

The Internet Protocol (IP) is a connectionless protocol that is responsible for addressing and routing packets. It breaks large packets down into smaller packets when they are travelling through a network that supports smaller packets. A *connectionless protocol means that a session is not established before data is exchanged*, and

Table 14.1 TCP layers

Layer	Description
Network interface layer	This layer is responsible for formatting packets and placing them on to the underlying network
Internet layer	This layer is responsible for network addressing. It includes the Internet protocol and the address resolution protocol
Transport layer	This layer is concerned with data transport and is implemented by TCP and the User Datagram Protocol (UDP)
Application layer	This layer is responsible for liaising between user applications and the transport layer
	It includes the File Transfer Protocol (FTP), telnet, Domain Name System (DNS) and Simple Mail Transfer Program (SMTP)

packet delivery with IP is not guaranteed as packets may be lost or delivered out of sequence. An acknowledgement is not sent when data is received, and the sender or receiver is not informed when a packet is lost or delivered out of sequence. The router forwards a packet only if it knows a route to the destination; otherwise the packet is dropped. Packets are dropped if their checksum is invalid or if their time to live is zero. The acknowledgement of packets is the responsibility of the TCP protocol. The ARPANET employed the TCP/IP protocols as a standard from 1983.

14.4 Birth of the Internet

The use of ARPANET was initially limited to academia and to the US military, and in the early years, there was little interest from industrial companies. It allowed messages to be sent between the universities that were part of ARPANET. There were over 2000 hosts on the TCP/IP enabled network by the mid-1980s.

It was decided to shut down the network by the late 1980s, and the National Science Foundation (NSF) commenced work on its successor, the NSFNET, in the mid-1980s. This network consisted of multiple regional networks connected to a major backbone. The original links in NSFNET were 56 Kbps, but these were updated to 1.544 Mbps T1 links in 1988. The NSFNET T1 backbone initially connected 13 sites, but this increased, as there was growing academic and industrial interest from around the world. The NSF quickly realized that the Internet had significant commercial potential.

The Internet began to become more international with nodes in Canada and several European countries. DARPA formed the Computer Emergency Response Team (CERT) to deal with any emergency incidents arising from the operation of the network.

The independent not-for-profit company, Advanced Network Services (ANS), was founded in 1991. It installed a new network (ANSNET) that replaced the NSFNET T1 network, and it operated over T3 (45Mbps) links. It was owned and operated by a private company rather than the US government, with the NSF focusing on the research aspects of networks rather than on the operational side.

The ANSNET network was a distributive network architecture operated by commercial providers such as Sprint, MCI and BBN. The various parts of the network were connected together by major network exchange points. These were termed Network Access Points (NAPs). There were over 160,000 hosts connected to the Internet by the late 1980s.

14.5 Birth of the World Wide Web

Tim Berners-Lee invented the World Wide Web, while working at CERN in 1990 [BL:00]. CERN is an important European centre for research in the nuclear field, and it is based in Switzerland. It employs several thousand physicists and scientists from around the world, and many visiting scientists spend a period of time there.

One of the problems that scientists at CERN faced in the late 1980s was keeping track of people, computers, documents and databases. The centre had many visiting scientists who spent several months there, as well as a large pool of permanent staff. There was no efficient and effective way in CERN at that time to share information among scientists.

A visiting scientist might need to obtain information or data from a CERN computer or to make the results of their research available to CERN. Berners-Lee came to CERN in the early 1980s, and he developed a program called 'Enquire' to assist with information sharing and in keeping track of the work of visiting scientists. He returned to CERN in the mid-1980s to work on other projects, and he devoted part of his free time to consider solutions to the information-sharing problem.

He built on several existing inventions such as the Internet; hypertext and the mouse. Ted Nelson invented hypertext in the 1960s, and it allowed links to be present in text. For example, a document such as a book contains a table of contents, an index and a bibliography. These are all links to material that is either within the book itself or external to the book. The reader of a book is able to follow the link to obtain the internal or external information. Doug Engelbart invented the mouse in the 1960s, and it allowed the cursor to be steered around the screen.

The major leap that Berners-Lee made was essentially a marriage of the Internet, hypertext and the mouse into what has become the World Wide Web. His vision and its subsequent realization benefited CERN and the wider world.

He created a system that gives every web page a standard address called the Universal Resource Locator (URL). Each page is accessible via the Hypertext Transfer Protocol (HTTP), and the page is formatted with the hypertext markup language (HTML). Each page is visible using a web browser. The key features of Berners-Lee invention are listed in Table 14.2.

Berners-Lee invented the well-known terms such as URL, HTML and World Wide Web, and he wrote the first browser program that allowed users to access web pages throughout the world. Browsers are used to connect to remote computers over the Internet and to request, retrieve and display the web pages on the local machine.

The early browsers included Gopher developed at the University of Minnesota and Mosaic developed at the University of Illinois. These were replaced in later

Table 14.2 Features of World Wide Web

Feature	Description
URL	Universal Resource Identifier (later renamed to Universal Resource Locator (URL)) provides a unique address code for each web page
HTML	Hypertext markup language (HTML) is used for designing the layout of web pages
HTTP	The Hypertext Transport Protocol (HTTP) allows a new web page to be accessed from the current page
Browser	A browser is a client program that allows a user to interact with the pages and information on the World Wide Web

years by Netscape, which dominated the browser market until Microsoft developed Internet Explorer. The development of the graphical browsers led to the commercialization of the World Wide Web.

The World Wide Web creates a space in which users can access information easily from any part of the world. This is done using only a web browser and simple web addresses. The user can then click on hyperlinks on web pages to access further relevant information that may be on an entirely different continent. Berners-Lee is now the director of the World Wide Web Consortium, and this MIT-based organization sets the software standards for the Web.

The invention of the World Wide Web was a revolutionary milestone in the history of computing. It transformed the use of the Internet from mainly academic use to where it is now an integral part of peoples' lives. Users may now surf the Web, i.e. hyperlink among the millions of computers in the world, and obtain information easily. It is revolutionary in that:

- No single organization is controlling the Web.
- No single computer is controlling the Web.
- Millions of computers are interconnected.
- It is an enormous marketplace of billions of users.
- The Web is not located in one physical location.
- The Web is a space and not a physical thing.

14.6 Applications of the World Wide Web

Berners-Lee realized that the World Wide Web offered the potential to conduct business in cyberspace, rather than the traditional way where buyers and sellers come together to do business in the marketplace.

> Anyone can trade with anyone else except that they do not have to go to the market square to do so

The growth of the World Wide Web has been phenomenal, and exponential growth rate curves became a feature of newly formed Internet companies and their business plans. The World Wide Web has been applied to many areas including:

- Travel industry (booking flights, train tickets, and hotels)
- E-marketing
- On-line shopping (e.g. www.amazon.com)
- Portal sites (such as Yahoo)
- Recruitment services
- Internet banking
- On-line casinos (for gambling)

Table 14.3 Characteristics of e-commerce

Feature	Description
Catalogue of products	The catalogue of products details the products available for sale and their prices
Well designed and easy to use	This is essential as otherwise the web site will not be used
Shopping carts	This is analogous to shopping carts in a supermarket
Security	Security of credit card information is a key concern for users of the Web, as users need to have confidence that their credit card details will remain secure
Payments	Once the user has completed the selection of purchases, there is a checkout facility to arrange for the purchase of the goods
Order fulfilment/order enquiry	Once payment has been received, the products must be delivered to the customer

- On-line auction sites (e.g. eBay)
- Newspapers and news channels
- Social media (Facebook and Twitter)

The prediction in the early days was that the new web-based economy would replace traditional bricks and mortar companies. It was expected that most business would be conducted over the Web, with traditional enterprises losing market share and going out of business. Exponential growth of e-commerce companies was predicted, and the size of the new web economy was estimated to be in trillions of US dollars.

New companies were formed to exploit the opportunities of the Web, and existing companies developed e-business and e-commerce strategies to adapt to the brave new world. Companies providing full e-commerce solutions were concerned with the selling of products or services over the Web to either businesses or consumers. These business models are referred to as business to business (B2B) or business to consumer (B2C). E-commerce web sites have the following characteristics (Table 14.3).

14.7 Dot-Com Companies

The success of the World Wide Web was phenomenal and it led to a boom in the formation of *new economy* businesses. These businesses were conducted over the Web and included the Internet portal company, Yahoo; the on-line book store, Amazon; and the on-line auction site, eBay. Yahoo provides news and a range of services, and most of its revenue comes from advertisements. Amazon initially sold books, but it now sells a collection of consumer and electronic goods. eBay brings buyers and sellers together in an on-line auction space.

Some of these new technology companies were successful and remain in business. Others were financial disasters due to poor business models, poor management and poor implementation of the new technology. Some of these technology companies offered an Internet version of a traditional bricks and mortar company, with others providing a unique business offering. For example, eBay offers an auctioneering Internet site to consumers worldwide which was a totally new service and quite distinct from traditional auctioneering.

David Filo and Jerry Yang founded Yahoo, and they used it to keep track of their personal interests and the corresponding web sites on the Internet. Filo and Yang were students at Stanford in California, and their list of interests grew over time and became too long and unwieldy. Therefore, they broke their interests into a set of categories and then subcategories, and this is the core concept of the web site.

There was a lot of interest in the site from other students, family and friends and a growing community of users. The founders realized that the site had commercial potential, and they incorporated it as a business in 1995. The company launched its initial public offering (IPO) 1 year later in April 1996, and it was valued at $850 million. Yahoo is a portal site and it offers free email accounts to users, a search engine, news, shopping, entertainment, health and so on. The company earns most of its revenue from advertisement (including the click through advertisements that appear on a Yahoo web page).

Jeff Bezos founded Amazon in 1995 as an on-line bookstore. Its product portfolio has expanded to include the sale of CDs, DVDs, toys, computer software and video games. Its initial focus was to build up the 'Amazon' brand throughout the world and to become the world's largest bookstore. It initially sold books at a loss by giving discounts to buyers in order to build market share. It was very effective in building its brand through advertisements, marketing and discounts.

It has become the largest on-line bookstore in the world and has a solid business model with a very large product catalogue, a well-designed web site with good searching facilities, good checkout facilities and good order fulfilment. It also developed an associate model, which allows its associates to receive a commission for purchases of Amazon products made through the associate site.

Pierre Omidyar founded eBay in 1995, and the site brings buyers and sellers together. Millions of items are listed, bought and sold on eBay every day. The sellers are individuals or international companies. Any legal product that does not violate the company's terms of service may be bought or sold on the site. A buyer makes a bid for a product or service and competes against several other bidders. The highest bid is successful, and payment and delivery is then arranged. The revenue earned by eBay includes fees to list a product and commission fees that are applied whenever a product is sold.

Any product listed that violates eBay's terms of service is removed from the site as soon as the company is aware of them. The company also has a fraud-prevention mechanism, which allows buyers and sellers to provide feedback on each other and to rate each other following the transaction. The feedback may be positive, negative or neutral, and relevant comments are included. This offers a way to help to reduce fraud as unscrupulous sellers or buyers will receive negative ratings and comments.

14.7.1 Dot-Com Failures

Several of the companies formed during the dot-com era were successful and remain in business today. Others had inappropriate business models or poor management and failed in a spectacular fashion. This section considers some of the dot-com failures and highlights the reasons for failure.

Webvan.com was an on-line grocery business based in California. It delivered products to a customer's home within a 30 min period of their choosing. The company expanded to several other cities before it went bankrupt in 2001. Many of its failings were due to management as the business model was reasonable, and today there are several successful on-line fresh food delivery businesses. The management was inexperienced in the supermarket or grocery business, and the company spent excessively on infrastructure. It had been advised to build up an infrastructure to deliver groceries as quickly as possible, rather than developing partnerships with existing supermarkets. It built warehouses and purchased a fleet of delivery vehicles and top of the range computer infrastructure before running out of money.

Ernst Malmsten and others founded Boo.com in 1998, as an on-line fashion retailer that was based in the United Kingdom. The company spent over $135 million of shareholder funds in less than 3 years, and it went bankrupt in 2000. Its web site was poorly designed for its target audience, and it went against many of the accepted usability conventions of the time. The web site was designed in the days before broadband, with 56 K modems used by most customers. However, its design included the latest Java and Flash technologies, and it took most users several minutes to load the first page of the web site. Further, the navigation of the web site was inconsistent and changed as the user moved around the site.

Other reasons for failure included poor management and leadership, lack of direction, lack of communication between departments, spirally costs left unchecked and crippling pay roll costs. Further, purchasers returned a large number of products, and there was no postage charge applied for this service. The company went bankrupt in 2000, and an account of its formation and collapse is in the book, *Boo Hoo*, [MaP:02]. This book is a software development horror story, and the poor software development practices employed are evident from the fact that the developers were working without any source code control mechanism in place. The net effect was that despite extensive advertising by the company, the users were not inclined to use the site.

Pets.com was an on-line pet supply company founded in 1998 by Greg McLemore. It sold pet accessories and supplies. It had a well-known advertisement as to *why one should shop at an on-line pet store*. The answer to this question was *because pets can't drive*! Its mascot (the Pets.com sock puppet) was well known. It launched its IPO in February 2000 just before the dot-com collapse.

Pets.com made investments in infrastructure such as warehousing and vehicles. It needed a critical mass of customers in order to break even and its management believed that it needed $300 million of revenue to achieve this. They expected that this would take a minimum of 4–5 years, and therefore there was a need to raise further capital. However, following the dot-com collapse, there was negative sentiment towards technology companies, and it was apparent that it would be unable to raise further capital. The management tried to sell the company without success, and it went into liquidation 9 months after its IPO.

Joseph Park and Yong Kang founded Kozmo.com in New York in 1998. It was an on-line company that promised free 1 h delivery of small consumer goods. It provided point-to-point delivery usually on a bicycle and did not charge a delivery fee. Its business model was deeply flawed, as it is expensive to offer point-to-point delivery of small goods within a 1 h period without charging a delivery fee. The company argued that they could make savings to offset the delivery costs, as they did not require retail space. It expanded into several cities in the United States and raised about $280 million from investors. The company ceased trading in 2001.

14.7.2 Business Models

A business model converts a business or technology idea into a commercial reality, and it needs to be appropriate for the company and its intended operating market. A company with an excellent business idea but with a weak business model may fail, whereas a company with an average business idea but an excellent business model may be quite successful. Several of the business models in the dot-com era were deeply flawed, and the eventual collapse of many of these companies was predictable. Chesbrough and Rosenbloom [ChR:02] have identified six key components in a business model (Table 14.4).

Table 14.4 Characteristics of business models

Constituent	Description
Value proposition	This describes how the product or service is a solution to a customer problem
Market segment	This describes the customers that will be targeted (including market segments)
Value chain structure	This describes where the company fits into the value chain [Por:98]
Revenue generation and margins	This describes how revenue will be generated, including revenue streams from sales, support, etc.
Position in value network	This involves identifying competitors and other players that can assist in delivering added value to the customer
Competitive strategy	This describes how it will develop a competitive advantage to be successful

14.7.3 Bubble and Burst

The initial public offering of Netscape in 1995 demonstrated the incredible value of the new Internet companies. The company had planned to issue the share price at $14, but it decided at the last minute to issue it at $28. The share price reached $75 later that day. This was followed by what became the dot-com bubble where there were a large number of public offerings of Internet stock and where the value of these stocks reached astronomical levels. Reality returned to the stock market when it crashed in April 2000, and share values returned to more realistic levels.

The vast majority of these Internet companies were losing substantial sums of money, and few expected to deliver profits in the short term. Financial instruments such as the balance sheet, profit and loss account and price to earnings ratio are normally employed to estimate the value of a company. However, investment bankers argued that there was a new paradigm in stock market valuation for Internet companies. This paradigm suggested that the potential future earnings of technology companies be considered in determining their value, and this was used to justify the high prices of shares, as frenzied investors rushed to buy these overpriced and overhyped stocks. Common sense seemed to play no role in decision-making. The dot-com bubble is characterized by:

- Irrational exuberance on the part of investors.
- Insatiable appetite for Internet stocks.
- Incredible greed from all parties involved.
- Following herd mentality.
- A lack of rationality and common sense by all concerned.
- Traditional method of company valuation not employed.
- Interest in making money rather than in building the business first.
- Questionable decisions by Federal Reserve chairman (Alan Greenspan).
- Questionable analysis by investment firms.
- Conflict of interest investment banks.
- Market had left reality behind.

There were winners and losers in the boom and collapse. Some investors made a lot of money from the bubble, with others including pension funds and life assurance funds making significant losses. The investment banks typically earned 5–7 % commission on each successful IPO, and it was not in their interest to question the boom too closely. Those who bought and disposed early obtained a good return, whereas those who kept their shares for too long suffered losses. The full extent of the boom can be seen in the rise and fall of the value of the Dow Jones and NASDAQ from 1995 to 2002.

The extraordinary rise of the Dow Jones (Fig. 14.2) from a level of 3800 in 1995 to 11,900 in 2000 represented a 200 % increase over 5 years or approximately 26 % annual growth (compound) during this period. The rise of the NASDAQ (Fig. 14.3) over this period is even more dramatic. It rose from a level of 751 in 1995 to 5000 in 2000, representing a 566 % increase during the period. This is equivalent to a 46 % compounded annual growth rate of the index.

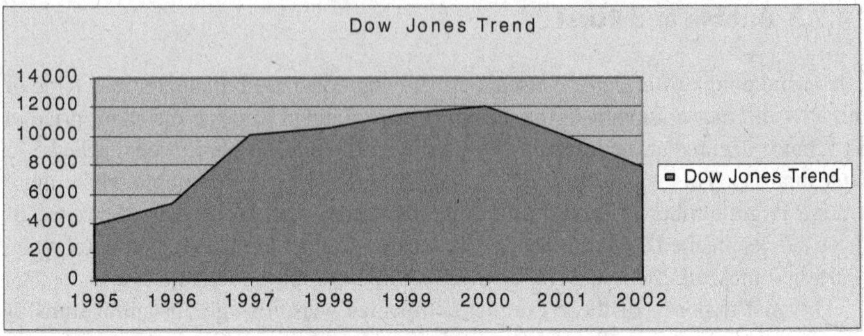

Fig. 14.2 Dow Jones (1995–2002)

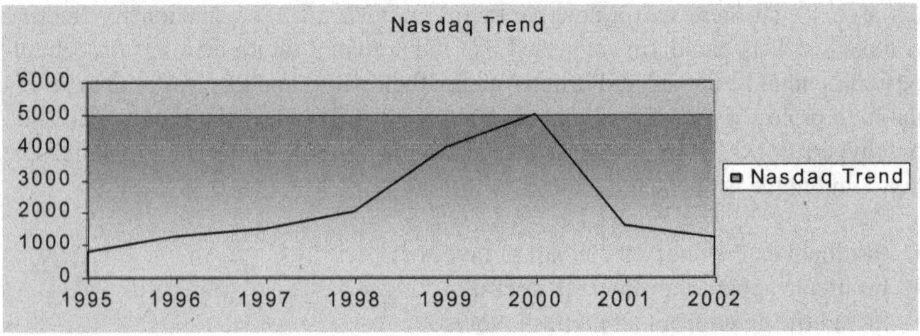

Fig. 14.3 NASDAQ (1995–2002)

The fall of the indices was equally as dramatic especially in the case of the NASDAQ. It peaked at 5000 in March 2000, and fell to 1200 (a 76 % drop) by September 2002. It had become clear that Internet companies were rapidly going through the cash raised at the IPOs, and analysts noted that a significant number would be out of cash by the end of 2000. Therefore, these companies would either go out of business or would need to go back to the market for further funding. This led to questioning of the hitherto relatively unquestioned business models of these Internet firms. Funding is easy to obtain when stock prices are rising at a rapid rate. However, when prices are static or falling, with negligible or negative business return to the investor, then funding dries up. The actions of the Federal Reserve in rising interest rates to prevent inflationary pressures also helped to correct the irrational exuberance of investors.

Some independent commentators had recognized the bubble, but their comments and analysis had been largely ignored. These included 'The Financial Times' and 'The Economist' as well as some commentators in the investment banks. Investors rarely queried the upbeat analysis coming from Wall Street and seemed to believe that rising stock prices would be a permanent feature of the US stock markets. Greenspan had argued that it is difficult to predict a bubble until after the event and that even if the bubble had been identified, it could not have been corrected without causing a contraction. Instead, the responsibility of the Fed according to Greenspan was to mitigate the fallout when it occurs.

There have, of course, been other stock market bubbles throughout history. For example, in the 1800s there was a rush on railway stock in England leading to a bubble and eventual burst of railway stock prices in the 1840s. There was a devastating property bubble and collapse (2002–2009) in the Republic of Ireland. The failure of the Irish political class, the Irish Central bank and financial regulators, the Irish Banking sector in their irresponsible lending policies and failures of the media in questioning the bubble are deeply disturbing. Its legacy remains, and while the country has made a remarkable recovery, the failures of so many at senior level in the state remain deeply disturbing.

14.8 E-Commerce Security

The World Wide Web consists of unknown users and suppliers with unpredictable behaviour operating in unknown countries around the world. These users and web sites may be friendly or hostile and the issue of trust arises:

- Is the other person whom they claim to be?
- Can the other person be relied upon to deliver the goods on payment?
- Can the other person be trusted not to inflict malicious damage?
- Is financial information kept confidential on the server?

Hostility may manifest itself in various acts of destruction. For example, malicious software may attempt to format the hard disk of the local machine, and if successful all local data will deleted. Other malicious software may attempt to steal confidential data from the local machine including bank account or credit card details. The *denial of service attack* is when a web site is overloaded by a malicious attack and where users are therefore unable to access the web site for an extended period of time.

The display of web pages on the local client machine may involve the downloading of programs from the server and running the program on the client machine. Standard HTML allows the static presentation of a web page, whereas many web pages include active content (e.g. Java applets or Active X). There is a danger that a Trojan Horse[5] may be activated during the execution of active content.

Security threats may be from anywhere (e.g. client side, server side, transmission) in an e-commerce environment, and therefore a holistic approach to security is required. Internal and external security measures need to be considered, with

[5] The origin of the term 'Trojan Horse' is from Homer's Iliad and concerns the Greek victory in the Trojan War. The Greek hero, Odysseus, and others hid in a wooden horse, while the other Greeks sailed away from Troy. This led the Trojans to believe that the Greeks had abandoned their attack and were returning to their homeland leaving behind a farewell gift for the citizens of Troy. The Trojans brought the wooden horse into the city, and later that night Odysseus and his companions opened the gates of Troy to the returning Greeks, and the mass slaughter of the citizens of Troy commenced, hence the phrase 'Bewared of Greeks bearing gifts'. Troy is located at the mouth of the Dardanelles in Turkey.

internal security generally implemented with good processes and procedures and assigning appropriate access privileges.

It is essential that the user is confident in the security provided as otherwise they will be reluctant to pass credit card details over the Web for purchases. Technologies such as Secure Sockets Layer (SSL) and Secure HTTP (S-HTTP) help to ensure security.

14.9 Review Questions

1. Describe the development of the Internet.
2. Describe the development of the World Wide Web and its key constituents.
3. Describe the applications of the World Wide Web.
4. Describe the key constituents of an electronic commerce site.
5. Describe a successful dot-com company that you are familiar with. What has made the company successful?
6. Describe a dot-com failure that you are familiar with. What caused the company to fail?
7. Discuss the key components of a business model.
8. Discuss security in an e-commerce environment.

14.10 Summary

This chapter considered the evolution of the Internet from the early work on packet switching and ARPANET to the subsequent development of the TCP/IP network protocols that specify how computers communicate and the standards for interconnecting networks and computers.

TCP/IP provides a transmission standard that deals with physical differences in host computers, routers and networks. It is designed to transfer data over networks which support different packet sizes and which may sometimes lose packets. TCP details how information is broken into packets and reassembled on delivery, whereas IP is focused on sending the packet across the network.

The invention of the World Wide Web by Tim Berners-Lee was a revolutionary milestone in computing. It transformed the Internet from mainly academic use to commercial use, and it led to a global market of consumers and suppliers. Today, the World Wide Web is an integral part of peoples' lives.

The growth of the World Wide Web was exponential, and the boom led to the formation of many 'new economy' businesses. These new companies conducted business over the Web as distinct from the traditional bricks and mortar companies. Some of these new companies were very successful (e.g. Amazon) and remain in business. Others were financial disasters due to poor business models, poor management and poor implementation of the new technology.

15 The Smartphone and Social Media

Abstract

A smartphone contains advanced computing capabilities that are attractive to users, and it arose as the outcome of the marriage of the existing mobile phone technology and PDA technology. A smartphone is more than a mobile device for making and receiving calls, and it is essentially a touch-based computer on a phone, which comes with its own keyboard, operating system, Internet access and third-party applications. It provides many other features such as a camera, maps, calendar, alarm clock and games. Today, the smartphone is ubiquitous.

We discuss the impact of Facebook and Twitter in social networking. Facebook is the leading social media site in the world, and it has become a way for young people to discuss their hopes and aspirations as well as a tool for social protest and revolution. Twitter has become a popular tool in political communication, and it is also an effective way for businesses to advertise its brand to its target audience.

Key Topics

PDA
Smartphone
Facebook
Tweets
Twitter

G. O'Regan, *Introduction to the History of Computing*, Undergraduate Topics in Computer Science, DOI 10.1007/978-3-319-33138-6_15

15.1 Introduction

Smartphones arose as the outcome of the marriage of the existing mobile phone technology and PDA technology, and they contain advanced computing capabilities that are attractive to users. Today, the smartphone is ubiquitous, with most people in advanced countries owning one.

We consider some of the events that led to the development of the smartphone, such as the introduction of the PDA by Apple and Palm. The introduction of the smartphone facilitated a major growth of social networking, as users were now able to communicate news events or update their personal information in real time. Social networking sites such as Facebook and Twitter have transformed human communication.

Social media involves the use of computer technology that allows the creation and exchange of user-generated content. These web-based technologies allow users to collaborate to discuss and modify user-created content. It has led to major changes in communication between individuals, communities and organizations.

Facebook helps users to keep in touch with friends and family, and it allows them to share their opinions on what is happening around the world. Users may upload photos and videos, express opinions and ideas and exchange messages, and Facebook allows their community of friends to be actively kept up to date on important events in their lives.

Facebook has become an important communication channel for educated young people to discuss their aspirations for the future, as well as their grievances with society and the state. It has even become an effective tool for protest and social revolution.

Twitter has become an effective way to communicate the latest news, and its effectiveness as a communication tool increases as the number of a person's followers grows. It allows a person or organization to determine what people are saying about it, including their positive or negative experiences. This allows direct interaction with the followers, and so it is a powerful way to engage the audience and to make people feel heard.

15.2 Evolution of the Smartphone

A smartphone is more than a mobile device for making and receiving calls, and it is essentially a touch-based computer on a phone, which comes with its own keyboard, operating system, Internet access and third-party applications. It provides many other features such as a camera, maps, calendar, alarm clock and games.

IBM (in a joint venture with BellSouth) introduced one of the earliest precursors of today's smartphones in 1993. This was the IBM Simon, and it included voice and data services. It acted as a mobile phone, a PDA and a fax machine, and it also included a touchscreen that could be used to dial numbers. It could send faxes and emails, as well as making or receiving calls, and included applications such as an address book, calendar and calculator. However it was an expensive and large bulky device, and it was priced at $900.

John Sculley, the CEO of Apple, coined the term *personal digital assistant*, and Apple introduced the first PDA, the Newton, in 1993. The Apple Newton included some nice features including limited handwriting recognition abilities. Xerox PARC had created a prototype PDA, the Dynabook, in the 1970s, but they did not commercialize it.

A PDA allows a large amount of data to be stored on a small handheld device. Palm introduced an early PDA device, the Palm Pilot 1000, which was used for mobile data, and it was introduced in 1996. It played an important role in popularizing the use of mobile data by business users. The Palm Pilot started the PDA industry, and it included 128Kb of memory and 16 MHz of processing power. It had better handwriting recognition capabilities than the Newton and a graphical user interface (GUI).

The Nokia 9000 Communicator was released in 1996, and this phone combined the features of a PDA and a mobile phone. It included a physical QWERTY keyboard, and it provided features such as email, calendar, address book and calculator. However, it did not provide the ability to browse the web, and a colour display was introduced in the Nokia 9210 in 1998.

Qualcomm introduced its pdQ smartphone in 1999, and this phone combined a Palm PDA with Internet connectivity capabilities. Research In Motion (RIM) released its first Blackberry devices in 1999, and these provided secure email communication into a single inbox. Samsung's first smartphone was the Samsung SPH-I300, which was released in 2001, and this Palm-powered smartphone is a distant ancestor of today's smartphones. Samsung introduced its SGH i607 smartphone in 2006, and this Windows-powered phone was inspired by Research in Motion's Blackberry phone.

Smartphone technology continued to evolve through the early 2000s, and Apple introduced its revolutionary *i*Phone in 2007. This Internet-based multimedia smartphone included a touchscreen and features such as a video camera, email, web browsing, text messaging and voice. The *i*Phone had a 3.5 inch 480×320 touchscreen, a QWERTY keyboard and 4GB of storage. Apple developed its own operating system, *i*OS, for the *i*Phone.

Google introduced its open-source Android operating system in the late 2007, and the first Android phone was introduced in the late 2008. Android is now the dominant operating system for smartphones and tablets, with *i*OS used on Apple's products. The Samsung Instinct was released in 2008, but it was based on an operating system developed by Samsung from various Java components. Although its touchscreen operating system was not in the same league as Apple's *i*OS, it became a competitor to Apple's *i*Phone.

Apple's *i*Phone 4 (Fig. 15.1) was introduced in 2010, and this powerful smartphone has a 3.5 inch 960×640 screen and a 5 megapixel camera. The Samsung Galaxy S smartphone was launched in 2010, and this touchscreen-enabled Android smartphone became extremely popular. The Samsung Galaxy S series of smartphones have been very successful and have become a major competitor to Apple's *i*Phone.

Fig. 15.1 Apple iPhone 4

Apple released the *i*Pad in 2010, which is a large screen tablet-like device that uses a touchscreen operating system. Samsung is a major competitor to Apple in the tablet market.

15.3 The Facebook Revolution

Facebook is the leading social networking site (SNS) in the world, and its mission is to make the world more open and connected. It helps users to keep in touch with friends and family, and it allows them to share their opinions on what is happening around the world. Users may upload photos and videos, express opinions and ideas and exchange messages. Facebook is very popular with advertisers as it allows them to easily reach a large target audience.

Mark Zuckerberg (Fig. 15.2) founded the company in 2004 while he was a student studying psychology at Harvard University. Zuckerberg was interested in programming, and he had already developed several social networking websites for his fellow students including *Facemash*, which could be used to rate the attractiveness of a person, and *CourseMatch* which allowed students to view people taking their degree.

Zuckerberg launched *The Facebook* (thefacebook.com) at Harvard in February 2004, and over a thousand Harvard students had registered on the site within the first 24 h. Over half of the Harvard student population had a profile on Facebook within the first month. The membership of the site was initially restricted to students at Harvard, then to students at the other universities in Boston, and then to students at the other universities in the United States. Its membership was extended to international universities from 2005.

The use of Facebook was extended beyond universities to anyone with an email address from 2006, and the number of registered users began to increase exponentially. The number of registered users reached 100 million in 2008 and 500 million in 2010 and exceeded one billion in 2012. It is now one of the most popular web sites in the world.

Fig. 15.2 Mark Zuckerberg

Facebook's business model is quite distinct from that of a traditional business in that it does not manufacture or sell any products. Instead it earns its revenue mainly from advertisements, and its business model is based on advertisement revenue, with advertisements targeted to its over 1.3 billion users based on their specific interests. Facebook is essentially selling its users to advertisers (i.e. the users are the product). The users really do all the work, and Facebook collects data about them (e.g. age, gender, location, education, work history and interests) and classifies and categorizes them, so that it is in a position to target advertisements that will potentially be of interest to them. This means that the advertisements are targeted to the right audience.

Social media have become important communication channels for educated young people to discuss their aspirations for the future, as well as their grievances with society and the state. The effectiveness of Facebook as a tool for protests and revolution is evident in the relatively short protests that culminated in the resignation of President Hosni Mubarak of Egypt in 2011.

Egypt has a young population with roughly 60 % of the population under the age of 30, and the country has faced many challenges since independence such as improving education and literacy for its young population, as well as finding jobs for its citizens.

Facebook provided a platform for Egyptian youth to discuss issues such as unemployment, low wages, police brutality and corruption. Young Egyptians set up groups on Facebook to discuss specific issues (e.g. a group that aimed to provide solidarity with striking workers was set up). Further momentum for revolution followed the beating and killing of Khaled Mohammed Said, as photos of his disfigured body were posted over the Internet and went viral. An influential Facebook group called *We Are All Khaled Said* was set up, and the killing provided a tangible focus for solidarity among young Egyptians.

The protests lasted for 18 days and it led to hundreds of thousands of young Egyptians taking to the streets and gathering in Tahrir Square in Cairo. They demanded an end to police brutality as well as the end of the 30-year reign of President Hosni Mubarak. The authorities reacted swiftly in closing down the Internet in Egypt, but this act of censorship failed to stop the protests against Mubarak. Social media played an important role in mobilizing protests and influencing the outcome of the revolution.

15.4 The Tweet

Twitter is a social communication tool that allows people to broadcast short messages. It is often described as the *SMS of the Internet*, and it is an online social media and microblogging site that allows its users to send and receive short 140-character messages called *tweets*. The restriction to 140 characters is to allow Twitter to be used on non-smartphone mobile devices. Twitter has over 300 million active users, and it is one of the most visited websites in the world. Users may access Twitter through its website interface, a mobile device app or SMS.

Jack Dorsey (Fig. 15.3) and others founded the company in 2006. Dorsey introduced the idea of an individual using an SMS service to communicate with a small group while he was still an undergraduate student at New York University. The word *twitter* was the chosen name for this new service, and its definition as *a short burst of information* and *chirps from birds* was highly appropriate.

Fig. 15.3 Jack Dorsey at the 2012 Time 100 Gala

Twitter messages are often about friends telling one another about their day, what they are doing, where they are and what they are thinking and doing, and Twitter has transformed the world of media, politics and business. It is possible to include links to web pages and other media as a tweet. News such as natural disasters, sports results and so on are often reported first by Twitter. The site has impacted political communication in a major way, as it allows politicians and their followers to debate and exchange political opinions. It allows celebrities to engage and stay in contact with their fans, and it provides a new way for businesses to advertise its brands to its target audience.

As a Twitter user, you select which other people who you wish to follow, and when you follow someone, their tweets show up in a list known as your *Twitter stream*. Similarly, anyone that chooses to follow you will see your tweets in their stream.

A *hashtag* is an easy way to find all the tweets about a particular topic of interest, and it may be used even if you are not following the people who are tweeting. It also allows you to contribute to the particular topic that is of interest. A hashtag consists of a short word or acronym preceded by the hash sign (#), and conferences, hot topics and so on often have a hashtag.

A word or topic that is tagged at a greater rate than other hashtags is said to be a *trending topic*, and a trending topic is often the result of an event that prompts people to discuss a particular topic. Trending may also result from the deliberate action of certain groups (e.g. in the entertainment industry) to raise the profile of a musician or celebrity and to market their work.

Twitter has evolved to become an effective way to communicate the latest news, and its effectiveness as a communication tool for an organization increases as the number of its followers grows. An organization may determine what people are saying about it, as well as their positive or negative experience in interacting with it. This allows the organization to directly interact with its followers, which is a powerful way to engage with its audience and to make people feel heard. It allows the organization to respond to any negative feedback and to deal with such feedback sensitively and appropriately.

The first version of Twitter was introduced in mid-2006, and it took the company some time to determine exactly what type of entity it actually was. There was nothing quite like it in existence, and initially it was considered a microblogging and social media site. Today it is considered to be an information network rather than just a social media site.

Twitter has experienced rapid growth from 400,000 tweets posted per quarter in 2007, to 100 million per quarter in 2008, to 65 million tweets per day from mid-2010, to 140 million tweets per day in 2011. Twitter's usage spikes during important events such as major sporting events, natural disasters, the death of a celebrity and so on. For such events, there may be over 100, 000 tweets per second.

Twitter's main source of revenue is advertisements through *promoted tweets* that appear in a user's timeline (Twitter stream). The first promoted tweets appeared from late 2011, and the use of a tweet for advertisement was ingenious. It helped to make the advertisement feel like part of Twitter, and it meant that an advertisement could go anywhere that a tweet could go. Advertisers are only charged when the user follows the links or retweets the original advertisements. Further, the use of tweets for advertisement meant that the transition to mobile was easy, and today about 80 % of Twitter use is on mobile devices.

Twitter has recently embarked on a strategy that goes beyond these advertisements to sell products directly (including to people who don't use Twitter). Twitter also earns revenue from a data licensing arrangement where it sells its information to companies who use this information to analyse consumer trends. Twitter analyses what users tweet in order to understand their intent. For more detailed information on Twitter, see [Sch:14].

15.5 Review Questions

1. What is a PDA?
2. What is a smartphone?
3. What is social media? Explain how sites such as Facebook and Twitter have transformed human communication.
4. Explain how a company may use social media to market new products to its customers.
5. Explain how social media has been used as a tool for protest and revolution.
6. Why has Twitter been described as the SMS of the Internet?

15.6 Summary

A smartphone is essentially touch-based computer on a phone, which comes with its own keyboard, operating system, Internet access and third-party applications. It provides many other attractive features such as a camera, maps, calendar, alarm clock and games. It arose from the marriage of mobile phone technology and PDA technology.

The smartphone has facilitated a major growth of social networking, as users are now able to communicate news or update their personal information in real time. Social media involves the use of computer technology that allows the creation and exchange of user-generated content. It has led to major changes in communication between individuals, communities and organizations. Social networking sites such as Facebook and Twitter have transformed human communication.

Facebook helps users to keep in touch with friends and family, and it allows them to share their opinions on what is happening around the world. Users may upload photos and videos, express opinions and ideas and exchange messages. It has become an important communication channel for young people to discuss their aspirations for the future, and it has also become an effective tool for mobilizing protests and social revolution.

Twitter has become an effective way to communicate the latest news, and its effectiveness as a communication tool increases as the number of its followers grows. It allows a person or organization to determine what people are saying about it, as well as their positive or negative experiences.

History of Programming Languages 16

Abstract
This chapter presents a short history of programming languages, starting with machine languages, to assembly languages, to early high-level procedural languages such as FORTRAN and COBOL, to later high-level languages such as Pascal and C and to object-oriented languages such as C++ and Java. Functional programming languages and logic programming languages are discussed, and there is a short discussion on the important area of syntax and semantics.

Key Topics
Generations of programming languages
Imperative languages
ALGOL
FORTRAN and COBOL
Pascal and C
Object-oriented languages
Java and C++
Functional programming languages
Logic programming languages
Syntax and semantics

16.1 Introduction

Hardware is physical and may be seen and touched, whereas software is intangible and is an intellectual undertaking by a team of programmers. Software is written in a particular programming language, and hundreds of languages have been

G. O'Regan, *Introduction to the History of Computing*, Undergraduate Topics in Computer Science, DOI 10.1007/978-3-319-33138-6_16

developed. Programming languages have evolved from the early days of computing with the earliest languages using machine code to instruct the computer. The next development was the use of assembly languages to represent machine language instructions. These were then translated into machine code by an assembler. The next step was to develop high-level programming languages such as FORTRAN and COBOL. These were easier to use than assembly languages and machine code and helped to improve quality and productivity.

A *first-generation* programming language (or 1GL) is a machine-level programming language that consists of 1 and 0 s. The main advantage of these languages is execution speed as they may be directly executed on the computer. These languages do not require a compiler or assembler to convert from a high-level language or assembly language into the machine code.

However, writing a program in machine code is difficult and error prone, as it involves writing a stream of binary numbers. This made the programming language difficult to learn and difficult to correct should any errors occur. The programming instructions were entered through the front panel switches of the computer system, and adding new code was difficult. Further, the machine code was not portable as the machine language for one computer could differ significantly from that of another computer. Often, the program needed to be totally rewritten for the new computer.

First-generation languages were used mainly on the early computers. A program written in a high-level programming language is generally translated by the compiler[1] into the machine language of the target computer for execution.

Second-generation languages, or 2GL, are low-level assembly languages that are specific to a particular computer and processor. However, assembly languages are easier to read than the first-generation machine code. They require considerably more programming effort than high-level programming languages and are more difficult to use for larger applications. The assembler converts the assembly code into the actual machine code to run on the computer. The assembly language is specific to a particular processor family and environment and is therefore not portable.

A program written in assembly language for a particular processor family needs to be rewritten for a different platform. However, since the assembly language is in the native language of the processor, it has significant speed advantages over high-level languages. Second-generation languages are still used today, but high-level programming languages have generally replaced them.

The *third-generation languages*, or 3GL, include high-level programming languages such as Pascal, C or FORTRAN. They are general-purpose languages and

[1]This is true of code generated by native compilers. Other compilers may compile the source code to the object code of a virtual machine, and the translator module of the virtual machine translates each byte code of the virtual machine to the corresponding native machine instruction. That is, the virtual machine translates each generalized machine instruction into a specific machine instruction (or instructions) that may then be executed by the processor on the target computer. Most computer languages such as C require a separate compiler for each computer platform (i.e. computer and operating system). However, a language such as Java comes with a virtual machine for each platform. This allows the source code statements in these programs to be compiled just once, and they will then run on any platform.

have been applied to business, scientific and general applications. They are designed to be easier for a human to understand and include features such as named variables, conditional statements, iterative statements, assignment statements and data structures. Early examples of third-generation languages are FORTRAN, ALGOL and COBOL. Later examples are languages such as C, C++ and Java. The advantages of these high-level languages are:

- Ease of readability
- Clearly defined syntax (and semantics[2])
- Suitable for business or scientific applications
- Machine independent
- Portability to other platforms
- Ease of debugging
- Execution speed

These languages are machine independent and may be compiled for different platforms. The early 3GLs were *procedural* in that they focus on how something is done rather than on what needs to be done. The later 3GLs were *object oriented*,[3] and the programming tasks were divided into objects. Objects may be employed to build larger programs, in a manner that is analogous to building a prefabricated building. Examples of modern object-oriented language are the Java language that is used to build web applications, C++ and Smalltalk.

High-level programming languages allow programmers to focus on problem-solving rather than on the low-level details associated with assembly languages. They are easier to debug and to maintain than assembly languages.

Fourth-generation languages specify what needs to be done rather than how it should be done. They are designed to reduce programming effort and include report generators and form generators. Report generators take a description of the data format and the report that is to be created and then automatically generate a program to produce the report. Form generators are used to generate programs to manage online interactions with the application system users. However, 4GLs are slow when compared to compiled languages.

A *fifth-generation* programming language, or 5GL, is a programming language that is based around solving problems using constraints applied to the program, rather than using an algorithm written by the programmer. Fifth-generation languages are designed to make the computer (rather than the programmer) solve the problem. The programmer specifies the problem and the constraints to be satisfied and is not concerned with the algorithm or implementation details. These languages are mainly used for research purposes especially in the field of artificial intelligence.

[2]The study of programming language semantics commenced in the 1960s. It includes work done by Hoare on axiomatic semantics, work done by Gordon Plotkin on operational semantics and work done by Scott and Strachey on denotational semantics.

[3]Norwegian Research originally developed object-oriented programming with their work on Simula 67 in the late 1960s.

Prolog is one of the best known fifth generation languages, and it is a logic programming language.

The task of deriving an efficient algorithm from a set of constraints for a particular problem is non-trivial, and to date this step has not been successfully automated. Fifth-generation languages are used mainly in academia.

16.2 Plankalkül

The earliest high-level programming language was Plankalkül developed by Konrad Zuse in 1946. It means 'Plan' and 'Kalkül' or, in other words, a calculus of programs. It is a relatively modern language for a language developed in 1946. There was no compiler for the language at the time, and it was only 50 years later that a compiler was finally developed for the language. The Free University of Berlin designed and developed a compiler in 2000, and the first Plankalkül program was run over 50 years after its conception.

The language employs data structures and Boolean algebra and includes a mechanism to define more powerful data structures. Zuse demonstrated that the Plankalkül language could be used to solve scientific and engineering problems, and he wrote several example programs including programs for sorting lists and searching a list for a particular entry. The main features of Plankalkül are:

- A high-level language.
- Fundamental data types are arrays and tuples of arrays.
- While construct for iteration.
- Conditionals are addressed using guarded commands.
- There is no GOTO statement.
- Programs are non-recursive functions.
- Type of a variable is specified when it is used.

The main constructs of the language are variable assignment, arithmetical and logical operations, guarded commands and while loops. There are also some list and set processing functions.

16.3 Imperative Programming Languages

Imperative programming is a programming style that describes computation in terms of a program state and statements that change the program state. The term *imperative* is a command to carry out a specific instruction or action. Similarly, imperative programming consists of a set of commands to be executed on the computer, and it is therefore concerned with *how* the program will be executed. The execution of an imperative command generally results in a change of state.

Imperative programming languages are quite distinct from *functional* and *logical programming languages*. Functional programming languages, like Miranda, have no global state, and programs consist of mathematical functions that have no side effects. In other words, there is no change of state, and the variable x will have the same value later in the program as it does earlier. Logical programming languages, like Prolog, define *what* is to be computed, rather than *how* the computation is to take place.

Most commercial programming languages are imperative languages, with interest in functional programming languages and relational programming languages being mainly academic. Imperative programs tend to be more difficult to reason about due to the change of state. Assembly languages and machine code are imperative languages.

High-level imperative languages use program variables and employ commands such as assignment statements, conditional commands, iterative commands and calls to procedures. An assignment statement performs an operation on information located in memory and stores the results in memory. The effect of an assignment statement is a change of the program state. A conditional statement allows a statement to be executed only if a specified condition is satisfied. Iterative statements allow a statement (or group of statements) to be executed a number of times.

High-level imperative languages allow the evaluation of complex expressions. These may consist of arithmetic operations and function evaluations, and the resulting value of the expression is assigned to memory.

FORTRAN was developed in the mid-1950s, and it was one of the earliest programming languages. ALGOL was developed in the late 1950s and 1960s, and it became a popular language for the expression of algorithms. COBOL was designed in the late 1950s as a programming language for business use. George Kemeny and Thomas Kurtz designed the BASIC (Beginner's All-purpose Symbolic Instruction Code) programming language in 1963. Niklaus Wirth developed Pascal in the early 1970s as a teaching language. Denis Ritchie at Bell Labs developed the C programming language in the early 1970s.

The Ada programming language was developed for the US military in the early 1980s. Object-oriented languages are imperative but include features to support objects. Bjarne Stroustrup designed C++ in 1985 as an object-oriented extension of the C language. Sun Microsystems released Java in 1996.

16.3.1 FORTRAN and COBOL

FORTRAN (FORmula TRANslator) was the first high-level programming language to be implemented. John Backus at IBM developed it in the mid-1950s, and the first compiler was available in 1957. The language includes named variables, complex expressions and subprograms. It was designed for scientific and engineering applications and remains the most important programming language for these domains. The main statements of the language include:

- Assignment statements (using the = symbol)
- IF statements
- GOTO statements
- DO loops

Fortran II was developed in 1958, and it introduced subprograms and functions to support procedural (or imperative) programming. Each procedure (or subroutine) contains computational steps to be carried out when it is called (at any point) during program execution. This could include calls by other procedures or by itself. However, recursion was not allowed until FORTRAN 90. FORTRAN 2003 provides support for object-oriented programming.

The basic types supported in FORTRAN include Boolean, integer and real. Support for double precision and complex numbers was added later. The language included relational operators for equality (.EQ.), less than (.LT.), and so on. FORTRAN is good at handling numbers and computation, and this is especially useful for mathematical and engineering problems. The following code (written in FORTRAN 77) gives a flavour of the language.

```
      PROGRAM HELLOWORLD
C     FORTRAN 77 SOURCE CODE COMMENTS FOR HELLOWORLD
         PRINT '(A)', 'HELLO WORLD'
         STOP
      END
```

FORTRAN remains a popular scientific programming language for application such as climate modelling, simulations of the solar system, modelling the trajectories of artificial satellites and simulation of automobile crash dynamics.

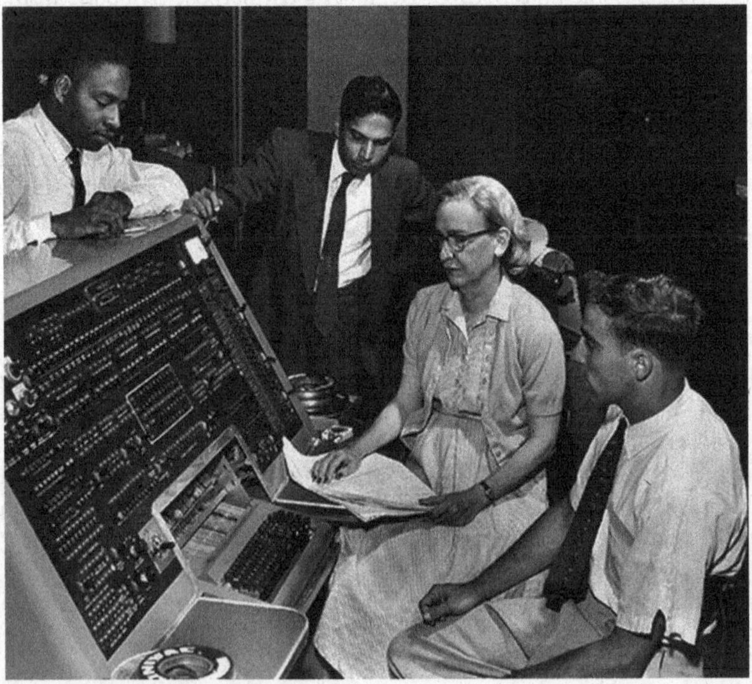

Fig. 16.1 Grace Murray and UNIVAC

It was initially weak at handling input and output, which was important to business computing. This led to the development of the COBOL programming language in the late 1950s.

The Common Business Oriented Language (COBOL) was the first business programming language, and it was introduced in 1959. Grace Murray Hopper[4] (Fig. 16.1) and a group of computer professionals called the Conference on Data Systems Languages (CODASYL) designed it with the objective of improving the readability of software source code. It has an English-like syntax designed to make it easy to learn the language. The only data types in the language were numbers and strings of text, and these may be grouped into arrays and records. The language is verbose:DIVIDE A BY B GIVING C REMAINDER D

COBOL was the first computer language whose use was mandated by the US Department of Defense. The language remains in use today, and there is an object-oriented version of the language.

16.3.2 ALGOL

ALGOL (ALGOrithmic Language) is a family of imperative programming languages, and it was originally developed in the mid-1950s and later revised in ALGOL 60 and ALGOL 68. It was designed to address some of the problems in FORTRAN, but it was not widely used. This may have been due to the refusal of IBM to support ALGOL and the dominance of IBM in the computing field.

A committee of American and European computer scientists designed the language, and it had a significant influence on later language design. ALGOL 60 [Nau:60] was the most popular member of the family, and Edsger Dijkstra developed an early ALGOL 60 compiler. John Backus and Peter Naur developed a method for describing the syntax of the ALGOL 58 programming language, which is known as Backus-Naur Form (or BNF).

ALGOL includes data structures and block structures. Block structures were designed to allow blocks of statements to be created (e.g. for procedures or functions). A variable defined within a block may be used within the block but is out of scope outside of the block.

ALGOL 60 introduced two ways of passing parameters to subprograms, and these are *call by value* and *call by name*. The call by value parameter passing technique involves the evaluation of the arguments of a function or procedure before the function or procedure is entered. The values of the arguments are passed to the function or procedure, and any changes to the arguments within the called function or procedure have no effect on the actual arguments. The call by name parameter passing technique is the default parameter passing technique in ALGOL 60. It involves re-evaluating the actual parameter expression each time the formal parameter is read. Call by name is used today in C/C++ macro expansion.

[4] Mary Hopper was a programmer on the Mark I, Mark II, Mark III and UNIVAC 1 computers. She was the technical advisor to the CODASYL committee.

ALGOL 60 includes conditional statements and iterative statements. It supports recursions: i.e. it allows a function or procedure to call itself. It includes:

- *Dynamic arrays.* These are arrays in which the subscript range is specified by variables.
- *Reserved words.* These are keywords that are not allowed to be used as identifiers by the programmer.
- *User-defined data types.* These allow the user to design their own data types.
- ALGOL uses bracketed statement blocks and it was the first language to use *begin-end* pairs for delimiting blocks.

ALGOL was used mainly by researchers in the United States and Europe. There was a lack of interest to its adoption by commercial companies due to the absence of standard input and output facilities in its description. ALGOL 60 became the standard for the publication of algorithms, and it had a major influence on later language development.

ALGOL evolved during the 1960s but not in the right direction. The ALGOL 68 committee decided on a very complex design rather than the simple and elegant ALGOL 60 specification. Tony Hoare remarked that:

ALGOL 60 was a great improvement on its successors.

16.3.3 Pascal and C

Niklaus Wirth designed the Pascal programming language in the early 1970s. It is named after Blaise Pascal (a seventeenth-century French mathematician), and it was based on the ALGOL programming language. It was intended as a language to teach students structured programming.

Structured programming [Dij:68] is concerned with rigorous techniques to design and develop programs, and there was intense debate on correct approaches to software development in the late 1960s. Dijkstra argued against the use of the GOTO statement 'GOTO Statement considered harmful' [Dij:68], and this influenced language design and led to several languages that did not include the GOTO statement.

The Pascal language includes constructs such as the conditional if statement; the iterative while, repeat and for statements; the assignment statement; and the case statement (which is a generalized if statement). The statement in the body of the repeat statement is executed at least once, whereas the statement within the body of a while statement may never be executed.

The language has several reserved words (known as keywords) that have a special meaning, and these may not be used as program identifiers. The Pascal program that displays 'Hello World' is given by:

```
program HELLOWORLD (OUTPUT);

begin
   WRITELN ('Hello, World!')
end.
```

Pascal includes several simple data types such as Boolean, integer, character and real. It also allows more advanced data types including arrays, enumeration types, ordinal types and pointer data types. It allows complex data types to be constructed from existing data types. Types are introduced by the reserved word 'type'.

```
type
   c = record
        a: integer;
        b: char
      end;
```

Pascal includes a 'pointer' data type, and this data type allows linked lists to be created by including a pointer type field in the record. The variable LINKLIST is a pointer to the data type B in the example below where B is a record:

```
type
  BPTR = ^B;
  B  = record
        A : integer;
        C : BPTR
      end;
var
  LINKLIST : BPTR;
```

Pascal is a block-structured language with programs structured into procedures and function blocks. These can be nested to any depth, and recursion is allowed. Each block has its own constants, types, variables and other procedures and functions, which are defined, within the scope of the block.

Pascal was criticized as being unsuitable for serious programming by Brian Kernighan and others [Ker:81]. Many of these deficiencies were addressed in later versions of the language. However, by then Denis Richie at Bell Labs had developed the C programming language, which became popular in industry. It is a general-purpose and a systems programming language.

It was originally designed as a language to write the kernel for the UNIX operating system. This was novel as operating systems were traditionally written in assembly languages. The success of C in writing the UNIX kernel led to its use on several other operating systems such as Windows and Linux. It also influenced later language development such as C++, and it is one of the most commonly used system programming languages. The language is described in detail in [KeR:78].

The language provides high-level and low-level capabilities, and a C program that is written in ANSI C with portability in mind may be compiled for a very wide variety of computer platforms and operating systems with minimal changes to the source code. The C language is now available on a wide range of platforms.

C is a procedural programming language and includes conditional statements such as the 'if statement', the 'switch statement', iterative statements such as the 'while' statement or 'do' statement and the assignment statement.

- If statement

```
        if (A == B)
            A = A+1;
        else
            A=A-1;
```
[5]

- Assignment statement

```
        i=i+1;
```

One of the first programs that people write in C is the Hello World program. This is given by:

```
main()

{
   printf("Hello, World\n");
}
```

It includes several predefined data types including integers and floating-point numbers.

- int (integer)
- long (long integer)
- float (floating-point real)
- double (double-precision real)

It allows more complex data types to be created using 'structs', which are similar to records in Pascal. It allows the use of pointers to access memory locations, which allows the memory locations to be directly referenced and modified. The result of the following example is to assign 5 to the variable x:

```
int  x;
int *ptr_x;

x=4;
ptr_x=&x;
*ptr_x=5;
```

C is a block-structured language, and a program is structured into functions (or blocks). Each function block contains its own variables and functions. A function may call itself (i.e. recursion is allowed).

[5]The semi-colon in Pascal is used as a statement separator, whereas it is used as a statement terminator in C.

One key criticism of C is that it is very easy to make errors in C programs and to thereby produce undesirable results. For example, one of the easiest mistakes to make is to accidentally write the assignment operator (=) for the equality operator (==). This totally changes the meaning of the original statement as can be seen below:

```
if (a == b)
    a++;        .... Program fragment A
else
    a--
if (a=b)
    a++;        .... Program fragment B
else
    a--
```

Both program fragments are syntactically correct and the intended meaning of a program is easily changed. The philosophy of C is to allow statements to be written as concisely as possible, and this is potentially dangerous.[6] The use of pointers potentially leads to problems as uninitialized pointers may point anywhere in memory and may therefore write anywhere in memory. Therefore, the effective use of C requires experienced programmers, well-documented source code and formal peer reviews of the source code by other developers.

16.4 Object-Oriented Languages

The traditional view of programming is that a program is a collection of functions or a list of instructions to be performed on the computer. *Object-oriented programming* is a paradigm shift in programming, where a computer program is considered to be a collection of objects that act on each other. Each object is capable of sending and receiving messages and processing data. That is, each object may be viewed as an independent entity or actor with a distinct role or responsibility.

An object is a *black box* which sends and receives *messages*. A black box consists of *code* (computer instructions) and *data* (information which these instructions operate on). The traditional way of programming kept code and data separate. For example, functions and data structures in the C programming language are not connected. However, in the object-oriented world, code and data are merged into a single indivisible thing called an *object*.

The reason that an object is called a black box is that the user of an object never needs to look inside the box, since all communication to it is done via messages. Messages define the *interface* to the object. Everything an object can do is represented by its message interface. Therefore, there is no need to know anything about what is in the black box (or object) in order to use it. The access to an object is only

[6] It is very easy to write incomprehensible code in C and even one line of C code can be incomprehensible. The maintenance of poorly written code is a challenge unless programmers follow good programming practice. This discipline needs to be enforced by formal reviews of the source code.

Table 16.1 Object-oriented paradigm

Feature	Description
Class	A class defines the abstract characteristics of a thing, including its attributes (or properties), and its behaviours (or methods). The members of a class are termed objects
Object	An object is a particular instance of a class with its own set of attributes. The set of values of the attributes of a particular object is called its state
Method	The methods associated with a class represent the behaviours of the objects in the class
Message passing	Message passing is the process by which an object sends data to another object or asks the other object to invoke a method
Inheritance	A class may have subclasses (or children classes) that are more specialized versions of the class. A subclass inherits the attributes and methods of the parent class. This allows the programmer to create new classes from existing classes. The derived classes inherit the methods and data structures of the parent class
Encapsulation (information hiding)	One fundamental principle of the object-oriented world is encapsulation (or information hiding). The internals of an object are kept private to the object and may not be accessed from outside the object. That is, encapsulation hides the details of how a particular class works and it requires a clearly specified interface around the services provided
Abstraction	Abstraction simplifies complexity by modelling classes and removing all unnecessary detail. All essential detail is represented, and non-essential information is ignored.
Polymorphism	Polymorphism is behaviour that varies depending on the class in which the behaviour is invoked. Two or more classes may react differently to the same message. The same name is given to methods in different subclasses, i.e. one interface, and multiple methods

through its messages, while keeping the internal details private. This is called *information hiding*[7] and is due to work by Parnas in the early 1970s.

The origins of object-oriented programming go back to the invention of Simula 67 at the Norwegian Computing Research Centre[8] in the late 1960s. It introduced the notion of a class and instances of a class.[9] Simula 67 influenced later languages such as the Smalltalk object-oriented language developed at Xerox PARC in the mid-1970s. Xerox introduced the term *object-oriented programming* for the use of objects and messages as the basis for computation. Most modern programming languages support object-oriented programming (e.g. Java and C++), and object-oriented features are added to many existing languages such as BASIC, FORTRAN and Ada. The main features of object-oriented languages are described in Table 16.1.

[7] Information hiding is a key contribution by Parnas to computer science. He has also done work on mathematical approaches to software quality using tabular expressions [ORg:06].

[8] The inventors of Simula 67 were Ole-Johan Dahl and Kristen Nygaard.

[9] Dahl and Nygaard were working on ship simulations and were attempting to address the huge number of combinations of different attributes from different types of ships. Their insight was to group the different types of ships into different classes of objects, with each class of objects being responsible for defining its own data and behaviour.

Object-oriented programming has become popular in large-scale software development, and it became the dominant paradigm in programming from the early 1990s. Its proponents argue that it is easier to learn and simpler to develop and maintain such programs. Its growth in popularity was helped by the rise in popularity of graphical user interfaces (GUI), which is well suited to object-oriented programming. The C++ programming language has become popular, and it is an object-oriented extension of the C programming language.

16.4.1 C++ and Java

Bjarne Stroustrup developed the C++ programming language in 1983 as an object-oriented extension of the C programming language. It was designed to use the power of object-oriented programming and to maintain the speed and portability of C. It provides a significant extension of C's capabilities, but it does not force the programmer to use the object-oriented features of the language.

A key difference between C++ and C is the concept of a class. A *class* is an extension to the C concept of a structure. The main difference is that while a C data structure can hold only data, a C++ class may hold both data and functions. An *object* is an instantiation of a class: i.e. the class is essentially the type, whereas the object is essentially a variable of that type. Classes are defined in C++ by using the keyword class:

```
class class_name
{
    access_specifier_1:
      member1;
    access_specifier_2:
      member2;
    ...
}
```

The members may be either data or function declarations, and an access specifier is included to specify the access rights for each member (e.g. private, public or protected). Private members of a class are accessible only by other members of the same class; public members are accessible from anywhere where the object is visible; protected members are accessible by other members of the same class and also from members of their derived classes. An example of a class in C++ is the definition of the class rectangle:

```
class CRectangle
{
  int x, y;
  public:
    void set_values (int,int);
    int area (void);
} rect;
```

Java is an *object-oriented programming language* developed by *James Gosling* and others at *Sun Microsystems* in the early 1990s. C and C++ influenced the syntax of the language, and the language was designed with portability in mind. The objective is for a program to be written once and executed anywhere. *Platform independence* is achieved by compiling the Java code into Java *bytecode, which* are simplified machine instructions specific to the Java platform.

This code is then run on a Java *virtual machine* (JVM) that interprets and executes the Java bytecode. The JVM is specific to the native code on the host hardware. The problem with interpreting bytecode is that it is slow compared to traditional compilation. However, Java has a number of techniques to address this including just in time compilation and dynamic recompilation. Java also provides automatic garbage collection. This is a very useful feature as it protects programmers who forget to deallocate memory (thereby causing memory leaks).

Java is a proprietary standard that is controlled through the Java Community Process. Sun Microsystems makes most of its Java implementations available without charge. The following is an example of the Hello World program written in Java:

```
class HelloWorld
{
  public static void main (String args[])
  {
      System.out.println ("Hello World!");
  }
}
```

16.5 Functional Programming Languages

Functional programming is quite distinct from imperative programming in that *it* involves the evaluation of *mathematical functions*. Imperative programming involves the execution of sequential (or iterative) commands that change the state. For example, the assignment statement alters the value of a variable, and the value of a given variable x may change during program execution.

There are no changes of state for functional programs. The fact that the value of x will always be the same makes it easier to reason about functional programs than imperative programs. Functional programming languages provide *referential transparency*: i.e. equals may be substituted for equals, and if two expressions have equal values, then one can be substituted for the other in any larger expression without affecting the result of the computation.

Functional programming languages use *higher-order* functions,[10] recursion, lazy and eager evaluation, monads[11] and *Hindley-Milner-type inference* systems.[12] These languages are mainly being used in academia, but there has been some industrial use, including the use of Erlang for concurrent applications in industry. Alonzo Church developed lambda calculus in the 1930s, and it provides an abstract framework for describing mathematical functions and their evaluation. It provides the foundation for functional programming languages. Church employed lambda calculus to prove that there is no solution to the *decision problem* for first-order arithmetic in 1936.

Lambda calculus uses transformation rules, and one of these rules is variable substitution. The original calculus developed by Church was untyped, but typed lambda calculi have since been developed. Any computable function can be expressed and evaluated using lambda calculus, but there is no general algorithm to determine whether two arbitrary lambda calculus expressions are equivalent. Lambda calculus influenced *functional programming languages* such as *LISP*, *ML* and *Haskell.*

Functional programming uses the notion of *higher-order functions*. Higher-order functions take other functions as arguments and may return functions as results. The derivative function $^{d}/_{dx}\, f(x) = f'(x)$ is a higher-order function. It takes a function as an argument and returns a function as a result. For example, the derivative of the function Sin(x) is given by Cos(x). Higher-order functions allow *currying which is* a technique developed by Schönfinkel. It allows a function with several arguments to be applied to each of its arguments one at a time, with each application returning a new (higher-order) function that accepts the next argument. This allows a function of n arguments to be treated as n applications of a function with one argument.

John McCarthy developed LISP at MIT in the late 1950s, and this language includes many of the features found in modern functional programming languages.[13] *Scheme* built upon the ideas in LISP. *Kenneth Iverson* developed *APL*[14] *in the early 1960s, and this language influenced Backus's FP programming language.* Robin Milner designed the ML programming language *in the early 1970s. David Turner* developed *Miranda in the mid-1980s.* The *Haskell programming language* was released in the late 1980s.

[10] Higher-order functions are functions that take functions as arguments or return a function as a result. They are known as operators (or functionals) in mathematics, and one example is the derivative function $^{dy}/_{dx}$ that takes a function as an argument and returns a function as a result.

[11] Monads are used in functional programming to express input and output operations without introducing side effects. The Haskell functional programming language makes use of uses this feature.

[12] This is the most common algorithm used to perform type inference. Type inference is concerned with determining the type of the value derived from the eventual evaluation of an expression.

[13] Lisp is a multi-paradigm language rather than a functional programming language.

[14] Iverson received the Turing Award in 1979 for his contributions to programming language and mathematical notation. The title of his Turing Award paper was 'Notation as a tool of thought'.

16.5.1 Miranda

Miranda was developed by David Turner at the University of Kent in the mid-1980s [Turn:85]. It is a non-strict functional programming language: i.e. the arguments to a function are not evaluated until they are actually required within the function being called. This is also known as lazy evaluation, and one of its main advantages is that it allows infinite data structures to be passed as an argument to a function. Miranda is a pure functional language in that there are no side effect features in the language. The language has been used for:

- Rapid prototyping
- Specification language
- Teaching language

A Miranda program is a collection of equations that define various functions and data structures. It is a strongly typed language with a terse notation.

```
z = sqr p / sqr q
    sqr k = k * k
    p = a + b
    q = a - b
    a = 10
    b = 5
```

The scope of a formal parameter (e.g. the parameter k above in the function sqr) is limited to the definition of the function in which it occurs.

One of the most common data structures used in Miranda is the list. The empty list is denoted by [], and an example of a list of integers is given by [1, 3, 4, 8]. Lists may be appended to by using the '++' operator. For example,

```
[1, 3, 5] ++ [2, 4] is [1, 3, 5, 2, 4].
```

The length of a list is given by the '#' operator:

```
# [1, 3] = 2
```

The infix operator ':' is employed to prefix an element to the front of a list. For example,

```
5 : [2, 4, 6] is equal to [5, 2, 4, 6]
```

The subscript operator '!' is employed for subscripting. For example,

```
Nums = [5,2,4,6]  then  Nums!0 is 5.
```

The elements of a list are required to be of the same type. A sequence of elements that contains mixed types is called a tuple. A tuple is written as follows:

```
Employee= ("Holmes", "222 Baker St. London", 50, "Detective")
```

A tuple is similar to a record in Pascal whereas lists are similar to arrays. Tuples cannot be subscripted but their elements may be extracted by pattern matching. Pattern matching is illustrated by the well-known example of the factorial function:

```
fac 0=1

fac (n+1) = (n+1) * fac n
```

The definition of the factorial function uses two equations, distinguished by the use of different patterns in the formal parameters. Another example of pattern matching is the reverse function on lists:

```
reverse [] = []

reverse (a:x) =reverse x ++ [a]
```

Miranda is a higher-order language, and it allows functions to be passed as parameters and returned as results. Currying is allowed and this allows a function of *n* arguments to be treated as *n* applications of a function with one argument. Function application is left associative: i.e. f x y means (f x) y. That is, the result of applying the function *f* to *x* is a function, and this function is then applied to *y*. Every function with two or more arguments in Miranda is a higher-order function.

16.5.2 Lambda Calculus

Lambda calculus (λ-calculus) was designed by *Alonzo Church* in the 1930s to study computability. It is a formal system that may be used to study function definition, function application, parameter passing and recursion. It may be employed to define what a *computable function* is, and any computable function may be expressed and evaluated using the calculus.

The lambda calculus is equivalent to the *Turing machine* formalism. However, lambda calculus emphasizes the use of transformation rules, whereas Turing machines are concerned with computability on primitive machines. Lambda calculus consists of a small set of rules:

Alpha-conversion rule (α-conversion)[15]
Beta-reduction rule (β-reduction)[16]
Eta-conversion (η-conversion)[17]

---

[15] This essentially expresses that the names of bound variables is unimportant.

[16] This essentially expresses the idea of function application.

[17] This essentially expresses the idea that two functions are equal if and only if they give the same results for all arguments.

Every expression in the λ-calculus stands for a function with a single argument. The argument of the function is itself a function with a single argument and so on. The definition of a function is anonymous in the calculus. For example, the function that adds one to its argument is usually defined as $f(x)=x+1$. However, in λ-calculus the function is defined as:

$$\lambda x.x+1 \qquad (\text{or equivalently as } \lambda z.z+1)$$

The name of the formal argument x is irrelevant and an equivalent definition of the function is $\lambda\, z.\ z+1$. The evaluation of a function f with respect to an argument (e.g. 3) is usually expressed by $f(3)$. In λ-calculus this would be written as $(\lambda\, x.\ x+1)$ 3, and this evaluates to $3+1=4$. Function application is *left associative*: i.e. $f\,x\,y=(f\,x)\,y$. A function of two variables is expressed in lambda calculus as a function of one argument, which returns a function of one argument. This is known as *currying* and has been discussed earlier. For example, the function $f(x, y)=x+y$ is written as $\lambda\, x.\ \lambda\, y.\ x+y$. This is often abbreviated to $\lambda\, x\, y.\ x+y$.

λ-Calculus is a simple mathematical system and its syntax is defined as follows:

```
<exp>::=<identifier>          |
    λ<identifier>.<exp> | --abstraction
    <exp><exp>          | --application
    (<exp>)
  -- Syntax of Lambda Calculus --
```

λ-Calculus's four lines of syntax plus *conversion* rules are sufficient to define B*ooleans*, *integers*, *data structures* and computations on them. It inspired LISP and modern functional programming languages.

16.6 Logic Programming Languages

Logic programming languages describe what is to be done, rather than how it should be done. These languages are concerned with the statement of the problem to be solved, rather than how the problem will be solved.

These languages use mathematical logic as a tool in the statement of the problem definition. Logic is a useful tool in developing a body of knowledge (or theory), and it allows rigorous mathematical deduction to derive further truths from the existing set of truths. The theory is built up from a small set of axioms or postulates and rules of inference derive further truths logically.

The objective of logic programming is to employ mathematical logic to assist with computer programming. Many problems are naturally expressed as a theory, and the statement of a problem to be solved is often equivalent to determining if a new hypothesis is consistent with an existing theory. Logic provides a rigorous way to determine this, as it includes a rigorous process for conducting proof.

Computation in logic programming is essentially logical deduction, and logic programming languages use first-order[18] predicate calculus. It employs theorem proving to derive a desired truth from an initial set of axioms. These proofs are constructive[19] in the sense that an actual object that satisfies the constraints is produced, rather than a reliance on a theoretical existence theorem. Logic programming specifies the objects, the relationships between them and the constraints that must be satisfied for the problem.

- The set of objects involved in the computation
- The relationships that hold between the objects
- The constraints of the particular problem

The language interpreter decides how to satisfy the particular constraints. *Artificial intelligence* influenced the development of logic programming, and *John McCarthy*[20] demonstrated that *mathematical logic* could be used for expressing knowledge. The first logic programming language was *Planner developed by Carl Hewitt at MIT in 1969. It uses a* procedural approach for knowledge representation rather than McCarthy's declarative approach.

The best-known logic programming languages is Prolog, which was developed in the early 1970s by *Alain Colmerauer* and *Robert Kowalski*. It stands for *pro*gramming in *log*ic. It is a goal-oriented language that is based on predicate logic. Prolog became an ISO standard in 1995. The language attempts to solve a goal by tackling the subgoals that the goal consists of:

$$\texttt{goal :- subgoal}_1\texttt{, …, subgoal}_n\texttt{.}$$

That is, in order to prove a particular goal, it is sufficient to prove $subgoal_1$ through subgoal*n*. Each line of a Prolog program consists of a rule or a fact, and the language specifies what exists rather than how. The following program fragment has one rule and two facts:

```
grandmother(G,S) :- parent(P,S), mother(G,P).
mother(sarah, isaac).
parent(isaac, jacob).
```

[18] First-order logic allows quantification over objects but not functions or relations. Higher-order logics allow quantification of functions and relations.

[19] For example, the statement $\exists x$ such that $x = \sqrt{4}$ states that there is an x such that x is the square root of 4, and the constructive existence yields that the answer is that $x = 2$ or x - -2, i.e. constructive existence provides more the truth of the statement of existence, and an actual object satisfying the existence criteria is explicitly produced.

[20] John McCarthy received the Turing Award in 1971 for his contributions to artificial intelligence. He also developed the programming language LISP.

The first line in the program fragment is a rule that states that G is the grandmother of S if there is a parent P of S and G is the mother of P. The next two statements are facts stating that isaac is a parent of jacob, and that sarah is the mother of isaac. A particular goal clause is true if all of its subclauses are true:

```
goalclause(Vg) :- clause1(V1),..,clausem(Vm)
```

A Horn clause consists of a goal clause and a set of clauses that must be proven separately. Prolog finds solutions by *unification:* i.e. by binding a variable to a value. For an implication to succeed, all goal variables V*g* on the left side of :- must find a solution by binding variables from the clauses which are activated on the right side. When all clauses are examined and all variables in V*g* are bound, the goal succeeds. But if a variable cannot be bound for a given clause, then that clause fails. Following the failure, Prolog *backtracks*, and this involves going back to the left to previous clauses to continue trying to unify with alternative bindings. Backtracking gives Prolog the ability to find multiple solutions to a given query or goal.

Most logic programming languages use a simple searching strategy to consider alternatives:

If a goal succeeds and there are more goals to achieve, then remember any untried alternatives and go on to the next goal.
If a goal is achieved and there are no more goals to achieve, then stop with success.
If a goal fails and there are alternative ways to solve it, then try the next one.
If a goal fails and there are no alternate ways to solve it, and there is a previous goal, then go back to the previous goal.
If a goal fails and there are no alternate ways to solve it, and no previous goal, then stop with failure.

Constraint programming is a programming paradigm where relations between variables can be stated in the form of constraints. Constraints specify the properties of the solution and differ from the imperative programming languages in that they do not specify the sequence of steps to execute.

16.7 Syntax and Semantics

There are two key parts to any programming language, namely, its syntax and semantics. The syntax is the grammar of the language, and a program needs to be syntactically correct with respect to its grammar. The semantics of the language is deeper and determines the meaning of what has been written by the programmer. The semantics of a language determines what a syntactically valid program will compute. A programming language is therefore given by:

```
Programming Language = Syntax + Semantics
```

The theory of the syntax of programming languages is well established, and Backus-Naur Form[21] (BNF) is employed to specify the grammar of languages. The grammar of a language may be input into a parser, which determines whether the program is syntactically correct. Chomsky[22] defined a hierarchy of grammars (regular, context-free, context sensitive). A BNF specification consists of a set of rules such as

```
<symbol>::=<expression with symbols>
```

where <symbol> is a *nonterminal* and the expression consists of sequences of symbols and/or sequences separated by the vertical bar '|' which indicates a choice. Symbols that never appear on a left side are called terminals. The partial definition of the syntax of various statements in a programming language is given below:

```
<loop statement>::=<while loop>|<for loop>
<while loop>::=while ()<statement>
<for loop>::= for ()<statement>
::=<assignment statement> |<loop statement>
<assignment statement>::=<variable>:=<expression>
```

The example above includes various nonterminals (<loop statement>, <while loop>, <for loop>, <condition>, <expression>, <statement>, <assignment statement> and <variable>). The terminals include 'while', 'for', ':=', '(' and ')'. The production rules for <condition> and <expression> are not included.

There are various types of grammars such as regular grammars, context-free grammars and context-sensitive grammars. A parser translates the grammar of a language into a parse table. Each type of grammar has its own parsing algorithm to determine whether a particular program is valid with respect to its grammar.

16.7.1 Programming Language Semantics

The formal semantics of a programming language is concerned with the meaning of programs. A program is written according to the rules of the language, and the compiler then checks that it is syntactically correct, and if so, it generates the equivalent machine code.[23]

The compiler must preserve the semantics of the language, and the syntax of the language gives no information as to the meaning of a program. It is possible to write syntactically correct programs that behave in quite a different way from the intentions of the programmer.

---

[21] Backus-Naur Form is named after John Backus and Peter Naur. It was created as part of the design of ALGOL 60 and used to define the syntax rules of the language.

[22] Chomsky made important contributions to linguistics and the theory of grammars. He is more widely known today as a critic of US foreign policy.

[23] Of course, what the programmer has written may not be what the programmer had intended.

Table 16.2 Programming language semantics

Approach	Description
Axiomatic semantics	Axiomatic semantics involves giving meaning to phrases of the language with logical axioms. This approach is based on mathematical logic, and it employs pre and post condition assertions to specify what happens when the statement executes. The relationship between the initial assertion and the final assertion essentially gives the semantics of the code
Operational semantics	The operational semantics for a programming language was developed by Gordon Plotkin [Plo:81]. It describes how a valid program is interpreted by a sequence of computational steps
	An abstract machine (SECD machine) may be defined to give meaning to phrases, by describing the transitions they induce on states of the machine
	A precise mathematical interpreter (such as the lambda calculus) may also give the semantics
Denotational semantics	Denotational semantics (originally called mathematical semantics) provides meaning to programs in terms of mathematical objects such as integers, tuples and functions
	Each phrase in the language is translated into a mathematical object that is the *denotation* of the phrase. Christopher Strachey and Dana Scott developed it in the mid-1960s

The formal semantics of a language is given by a mathematical model, which describes the possible computations described by the language. The three main approaches to programming language semantic are axiomatic semantics, operational semantics and denotational semantics. A short summary of each approach is described in Table 16.2, and more detailed information is in [ORg:06, ORg:12].

16.8 Review Questions

1. Describe the five generations of programming languages.
2. Describe the early use of machine code.
3. Describe the early use of assembly languages.
4. Describe the key features of Fortran and COBOL.
5. Describe the key features of Pascal and C.
6. Discuss the key features of object-oriented languages.
7. Explain the differences between imperative programming languages and functional programming languages.
8. What are the key features of logic programming languages?
9. What is the difference between syntax and semantics?
10. Explain the main approaches to programming language semantics.

16.9 Summary

This chapter considered the evolution of programming languages from the older machine languages, to the low-level assembly languages, to high-level programming languages and object-oriented languages, to functional and logic programming languages. Finally, the syntax and semantics of programming languages were briefly discussed.

The advantages of the machine languages are execution speed and efficiency. It is difficult to write programs in these languages, as the program involves a stream of binary numbers. These languages were not portable, as the machine language for one computer may differ significantly from the machine language of another.

The second-generation languages, or 2GLs, are low-level assembly languages that are specific to a particular computer and processor. These are easier to write and understand, but they must be converted into the actual machine code to run on the computer. The assembly language is specific to a particular processor family and environment and is therefore not portable. However, their advantages are execution speed, as the assembly language is the native language of the processor.

The third-generation languages, or 3GLs, are high-level programming languages. They are general-purpose languages and have been applied to business, scientific and general applications. They are designed to be easier to understand and to allow the programmer to focus on problem solving. Their advantages include ease of readability and portability and ease of debugging and maintenance. The early 3GLs were procedure oriented and the later 3GLs were object oriented.

Fourth-generation languages, or 4GLs, are languages that consist of statements similar to human language. Most fourth-generation languages are non-procedural and are often used in database programming. They specify what needs to be done rather than how it should be done, and they have been used as report generators and form generators.

Fifth-generation programming languages or 5GLs, are programming languages that is based around solving problems using logic programming or applying constraints to the program. They are designed to make the computer (rather than the programmer) solve the problem. The programmer only needs to be concerned with the specification of the problem and the constraints to be satisfied and does not need to be concerned with the algorithm or implementation details.

History of Operating Systems 17

Abstract
This chapter presents a short history of operating systems including the IBM OS/360, which was the operating system for the IBM System/360 family of computers. We discuss the MVS and VM operating systems, which were used on the IBM System/370 mainframe computer. Ken Thompson and Dennis Ritchie developed the popular UNIX operating system at Bell Labs in the early 1970s. This is a multi-user and multitasking operating system and was written almost entirely in C. DEC developed the VAX/VMS operating system in the late 1970s for its VAX family of minicomputers. Microsoft developed MS/DOS for the IBM personal computer in 1981, and it introduced Windows as a response to the Apple Macintosh. There is a short discussion on Android and iOS, which are popular operating systems for mobile devices.

Key Topics
MVS
VM
OS/360
UNIX
MS/DOS
Windows
Android
iOS

G. O'Regan, *Introduction to the History of Computing*, Undergraduate Topics in Computer Science, DOI 10.1007/978-3-319-33138-6_17

17.1 Introduction

An *operating system* is a collection of software programs that controls the hardware of a computer and makes it usable. It makes the computing power of the hardware available to the users of the computer, and it manages the hardware to achieve good system performance. An operating system manages system hardware such as the processors, storage, input/output devices, communication devices and data, and it provides functionality such as sharing hardware among users, scheduling resources among users, preventing users from interfering with each other, facilitating input/output, recovering from errors and handling network communication.

The earliest computers did not have an operating system, and the user had exclusive control over a large computer for a specified period of time. The user entered the program one bit at a time in machine code (initially using mechanical switches and later with a stack of punched cards) and waited for the results. People began to develop libraries to share code for common activities, and these are in a sense the precursor of today's operating systems.

The earliest operating systems were designed in the 1950s with the goal of making more efficient use of expensive computer resources. These batch-processing systems ran one job at a time, and programs and data were submitted in groups (or batches).

These evolved during the early 1960s into multi-batch systems that were designed to improve utilization of the expensive computer resources. They could handle several diverse jobs at once, and running several jobs offered a way to optimize computer utilization. One job could be using the processor while another job could be using the various I/O devices. These later batch-processing systems contained many peripheral devices such as card readers, card punches, printers, tape drives and disk drives. Jobs were normally submitted on punched cards and computer tape, and often a user's job could sit for hours (days) on an input table until it was processed. However, even a very slight error in a program would cause the program to fail, and it would require resubmission. This meant that software development in this environment was very slow. This led operating system designers to develop the concept of multiprogramming, in which several jobs are in main memory at once, and the concept of interrupts, where an interrupt allows one unit to gain the attention of another, and the state of the interrupted unit is saved prior to the processing of the interrupt and restored once processing is complete.

MIT developed the CTSS time-sharing system in the early 1960s, and this operating system provided users with typewriter-like terminals to obtain computing power from the machine. CTSS ran a conventional batch stream (to ensure high utilization of expensive computer resources), but it was also able to give fast responses to users who were editing or debugging programs. It was a highly interactive environment where the computer provided rapid responses to user requests. IBM began work on the CP/CMS operating system in 1964, and this would eventually evolve into IBM's VM operating system.

IBM announced the System/360 family of computers in 1964, and the computers in the family were designed to use the IBM System/360 operating system (OS/360).

OS/360 was a batch-oriented operating system, and IBM supported three variants of OS/360, which allowed multiprogramming for mid-range and top-range members of the family. The other major operating system used in the System/360 was the Disk Operating System (DOS/360).[1] The IBM System/360 evolved over time into the System/370 series.

MIT's successor to the CTSS operating system was a general time-sharing operating system called *Multics*, and Bell Labs was initially involved in its development. UNIX arose out of work on the development of Multics, and it was developed at Bell Labs in the early 1970s. It is a multitasking and multi-user operating system.

The IBM PC was introduced in 1981, and IBM outsourced the development of the operating system to a small company called Microsoft. The terms of the deal with IBM allowed Microsoft the right to license its operating system, MS/DOS, on IBM compatibles, with PC/DOS (or simply DOS) reserved for IBM personal computers only. MS/DOS managed floppy disks and files, input and output and memory, and it contained an external command processor that interpreted user commands and allowed the user to interact with the system.

The Macintosh was a paradigm shift for the computer industry when it was introduced in 1984. Its MAC operating system was GUI based, friendly, intuitive and easy to use, and it was clear that the future of operating systems was in GUI-driven systems, rather than primitive command-driven operating systems such as MS/DOS.

Microsoft Windows is a family of graphical operating systems developed by Microsoft, and it was Microsoft's initial response to Apple's GUI operating system. Windows has evolved to become the dominant operating system on laptops and personal computers, but it has failed to make an impact on the smartphone operating system market, which is dominated by Apple's iOS and Google's Android operating systems.

The Android operating system was designed mainly for touchscreen smartphones and tablets, and it was developed by Google and the Open Handset alliance. Android is built on the Linux kernel, and its first version was released in late 2007.

The iOS operating system is a mobile operating system employed on Apple's mobile devices such as smartphones and tablets. It was introduced in 2007. For more detailed information on operating systems, see [AnDa:14, Dei:90].

17.2 OS/360

IBM announced the System/360 family of computers in 1964, and the family of computers was designed to use the IBM System/360 operating system (OS/360). OS/360 was a batch-oriented operating system, and IBM supported three variants of it. These were OS/360 PCP (Principal Control Program), OS/360 MFT (Multiple Programming with a Fixed number of Tasks) and OS/360 MVT (Multiple Programming with a Variable number of Tasks).

[1] Not to be confused with DOS used on IBM personal computers.

OS/360 PCP was the simplest version, and it could run only one program at a time. The smaller members of the System/360 family used it. OS/360 MFT could run several programs at once, but only after partitioning the memory required to run each. It was subject to the limitation that if a program was idle, its allocated memory was unavailable to other programs. It was developed as an interim solution pending the delayed introduction of OS/360 MVT. However, the simpler MFT continued in use for many years due to problems with MVT.

OS/360 MVT was the most sophisticated version of OS/360, and it was intended for the largest members in the System family. It allowed memory divisions to be recreated as needed, and it was able to allocate all of a computer's memory (if required) to a single large job. Further, whenever memory was available, OS/360 MVT searched a queue of jobs to see if any could be run on the available memory. OS/360 MVT was introduced in 1967.

All three versions of OS/360 provided similar features from the point of view of application programs. This included the same Application Programming Interface (API), the same job control language (JCL) for initiating batch jobs, the same access methods for reading and writing files and data communication, the same spooling facility and multitasking.

OS/360 MVT evolved over time to become OS/VS2 following the introduction of virtual memory in the IBM System/370. OS/VS2 was later renamed to OS/MVS.

17.3 MVS

IBM introduced the Multiple Virtual Storage (MVS) operating system in 1974, and it was an enhancement of the MVT version of the OS/360 operating system that supported virtual memory. It was the most commonly used operating system on the IBM System/370 and System/390 mainframe computers.

The System/370 was an enhancement of the System/360 architecture in that it provided virtual storage capabilities, where *virtual storage* allows a much larger storage space to be addressed than is available in the primary memory of the computer. The concept of virtual storage dates back to the design of the Atlas Computer at the University of Manchester in 1960, and the two most common methods of implementing virtual storage are paging and segmentation.

The 24-bit addressing of the System/370 meant that each user (or job) had a 16-megabyte (2^{24}) virtual address space (i.e. 256 segments, with each segment containing 16 pages, and each page contained 4096 bytes).

MVS provides multiprogramming and multiprocessing capabilities, and it is a large operating system designed with performance, reliability and availability in mind. The operating system has recovery routines that gain control in the event of an operating system failure, and it attempts recovery from hardware errors.

MVS includes a master scheduler that initializes the system and responds to commands issued by the system operator. It contains a job entry subsystem that allows jobs to be entered into the system. Its system management facility collects information to account for system use and to analyse system performance.

Its time-sharing option (TSO) provides users with interactive editing, testing and debugging capabilities. Its data management functionality handles all input/output and file management activities. Its telecommunication functionality allows remote terminal users to access MVS.

17.4 VM

The virtual machine (VM) operating system makes a single machine appear as several real machines (Fig. 17.1). The user at a VM virtual machine sees the equivalent of a complete real machine, even though it is an illusion and just appears to be a real machine to the user. A virtual machine runs programs in a similar way to a real machine, and the user communicates with the virtual machine through a terminal. The most widely used virtual machine operating system is IBM's VM, and it is used on an IBM System/370 mainframe. It created the illusion that each user operating at a terminal had access to a complete IBM System/370, including the input/output devices.

VM can run several different operating systems at once, each of them on its own virtual machine. This is a very attractive feature as running multiple operating systems offers a form of backup in the event of failure. The operating systems running on virtual machines perform their normal functions such as storage management, control of input/output, processor scheduling and multiprogramming. Virtual machines create virtual processors, virtual storage and virtual I/O devices. The VM user may run operating systems such as MVS, VM/370, AIX/370 or VM itself.

The main components of VM are the Control Program (CP), the Conversational Monitor System (CMS), the Remote Spooling Communications Subsystem (RSCS), the Interactive Problem Control System (IPCS) and the CMS batch.

CP creates the environment in which virtual machines may execute, and it provides support for the various operating systems that may be used to control the IBM System/370. It manages the real machine underlying the virtual machine environment and gives each user access to the facilities of the real machine. CMS is an applications system with editors, debugging tools and various application packages.

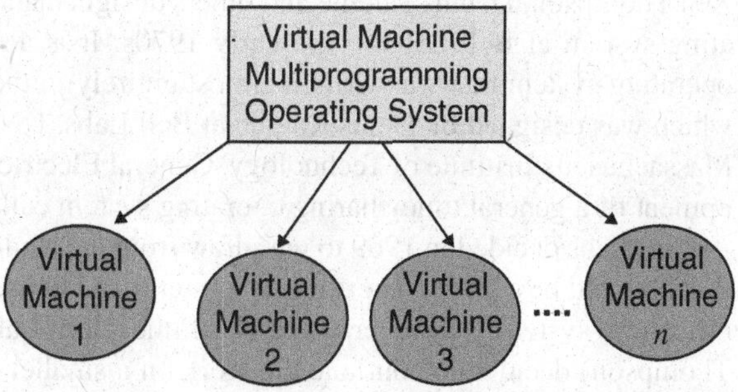

Fig. 17.1 Virtual machine operating system

RSCS provides the ability to transmit and receive files, and IPCS is used for on-line analysis and for fixing VM software problems. The CMS batch facility allows the user to submit longer jobs for batch processing.

17.5 VMS

The VAX Virtual Memory System (VMS) was designed as the operating system for the VAX family of minicomputers. Digital Equipment Corporation (DEC) introduced it in the late 1970s, and DEC was a major player in the minicomputer market with its popular family of PDP and VAX minicomputers. The models in the VAX family of computers all had the same architecture, and they could all run the VMS operating system.

David Cutler and others at DEC designed VMS as a high-end, secure, scalable, multi-user, multitasking and virtual memory operating system, and it supported a broad class of applications and systems. DEC developed VAX and VMS together, and the designers balanced the trade-offs between the work done by the hardware and the work done by the operating system.

VAXes may operate together in a peer-to-peer relationship, where any VAX may be a client or any may be a server. This allows flexibility when several computers perform tasks in cooperation. Several VAXes may be connected together so that they work as a cooperating unit called a VAXcluster.

VMS expanded the memory of the machine by disk or other peripheral storage to act as extra memory. The VAX-11 provided a 32-bit virtual address space per process, divided into 512 byte pages. VMS used paging and segmentation, with the first 23 bits used as the virtual page number (VPN), and a 9-bit offset within the page.

VMS was a popular and easy to use operating system. Its commands are easy to remember English-like words, and it has an extensive on-line help system. It included utilities such as a mail program and a text editor. Open VMS is the latest version of the operating system and is sold by HP.

17.6 UNIX

Ken Thompson, Dennis Ritchie and others designed and developed the UNIX operating system at Bell Labs in the early 1970s. It is a multitasking and multi-user operating system that was written almost entirely in the C programming language, which was designed by Denis Ritchie at Bell Labs. UNIX arose out of work by the Massachusetts Institute of Technology, General Electric and Bell Labs on the development of a general time-sharing operating system called *Multics*.

Bell Labs decided in 1969 to withdraw from the Multics project (as they believed that it would be a large and expensive system) and to use General Electric's GECOS operating system. However, several of the Bell Lab researchers (led by Ken Thompson) decided to continue the work on a smaller-scale operating system, and the name 'UNIX' was coined by Brian Kernighan. The first version of UNIX was written on a digital PDP-7 minicomputer in assembly language, and Dennis Ritchie joined the project. He helped in rewriting UNIX in the C programming language for the PDP-11 computer in 1973, which had recently been introduced. Thompson and Ritchie later received the Turing Award for their design and development of the UNIX operating system. Microsoft introduced XENIX, a commercial version of UNIX, in 1980.

The use of C helped to make UNIX more portable, and it became a widely used operating system. Universities and the US government used it initially, but it later became popular in industry. It is a powerful and flexible operating system, and it is used on a variety of machines from micros to supercomputers. It is designed to allow several programmers to access the computer at the same time, and to share its resources, and it offers powerful real-time sharing of resources.

It includes features such as *multitasking* which allows the computer to do several things at once, *multi-user* capability which allows several users to use the computer at the same time and *portability* of the operating system which allows it to be used on several computer platforms with minimal changes to the code. It includes a collection of tools and applications. There are three levels of the UNIX system: *kernel, shell* and *tools and applications*.

The kernel is the central part of the UNIX operating system, and it provides systems services to applications programs. This includes services for process management, memory management and input/output management. UNIX manages many concurrent processes.

The UNIX shell is a command interpreter that acts as the interface between the user and the operating system. There are a number of popular shells for UNIX including the Bourne shell and Korn shell. UNIX uses a hierarchical file system with the root node at its origin, with each directory entry containing files and other directories. For a more detailed account of UNIX, see [Rob:05].

17.7 MS/DOS

We discussed the introduction of the IBM personal computer in an earlier chapter, as well as the controversy with respect to the development of the PC/DOS operating system for the IBM PC. Digital Research, the developers of the CP/M operating system, lost out on the major opportunity of supplying the operating system for the IBM PC, and instead it was Microsoft that reaped the benefits. The terms of the deal with IBM allowed Microsoft the right to license its operating system, MS/DOS, on IBM compatibles, whereas PC/DOS (or simply DOS) was reserved for IBM personal computers only.

The IBM PC was introduced in 1981, and the first version of the operating system was compatible with Digital Research's CP/M operating system (as it essentially was CP/M). It managed floppy disks and files, input and output and memory, and it contained an external command processor that interpreted user commands and allowed the user to interact with the system.

MS/DOS version 2.0 was introduced in 1983 and it was designed to support the 10 MB hard disk on the IBM PC/XT, as well as providing support for device drivers. Microsoft had previously licensed XENIX (their commercial version of UNIX) from AT&T, and MS/DOS 2.0 was a move towards XENIX. It employed a hierarchical file system, and a unique pathname identified each file (similar to XENIX). It provided limited multitasking for background print spooling. The hard disk on the XT helped to establish the IBM PC in the business marketplace.

The open architecture of the IBM PC led to the development of cheaper IBM-compatible personal computers (clones of the IBM PC but cheaper), and they rapidly gained market share, as it was difficult for IBM to compete on price. This led to massive international demand for MS/DOS (which was the operating system for IBM compatibles and clones).

MS/DOS 3.0 was released in 1984 and it provided support for the IBM PC/AT, which had a 20 MB hard disk. Several versions of MS/DOS followed through the 1980s and 1990s and were used with Microsoft Windows 95 and Windows Millennium. Today, Microsoft Windows is the operating system used on personal computers, and MS/DOS is now of historical interest.

17.8 Microsoft Windows

Microsoft Windows is a family of graphical operating systems developed by Microsoft. The original Windows 1.0 operating environment was introduced in late 1985 as a graphical operating system shell for its command-driven MS/DOS operating system. It was Microsoft's initial response to Apple's GUI operating system.

The Apple Macintosh was released in 1984, and its MAC operating system was GUI based and a paradigm shift for the computer industry. It was friendly, intuitive and easy to use, and it was clear that the future of operating systems was in GUI-driven systems, rather than primitive command-driven operating systems such as MS/DOS.

The early versions of Windows were not complete operating systems as such and were instead graphical shells in that they ran on top of MS/DOS and extended the operating system. Windows 1.0 used MS/DOS for file system services, and it also included applications such as a calculator, calendar and clock. However, Windows differed from MS/DOS in that it allowed multiple graphical applications to be run at the same time, and this was done through cooperative multitasking.

Windows 2.0 was introduced in 1987 and it was more popular than its predecessor. It included improvements to the user interface and to memory management. Windows 3.0 improved the design of the operating system, and it used virtual memory and virtual device drivers that allowed arbitrary devices to be shared between multitasked DOS applications. It was introduced in 1990, and it was the first Windows operating system to achieve commercial success.

Windows 3.1 was introduced in 1992, Windows 95 in 1995, Windows 98 in 1998 and Windows Millennium (ME) in 2000. Windows ME provided expanded multimedia capabilities including the Windows Media Player, and it was the last

DOS-based version of Windows. Windows ME was criticized for its speed and instability.

Windows XP was introduced in 2001 and it was marketed into a 'Home' edition for personal users and a 'Professional' edition for business users. Windows Vista was released in 2006, Windows 7 in 2009, Windows 8 in 2012 and Windows 10 in 2015.

Microsoft Windows dominates the personal computer and laptop market with over 90 % market share. Windows has not been as successful on mobile computing platforms such as mobile phones and tablets, where Google's Android operating system is the dominant platform.

17.9 Mobile Operating Systems

Android (Fig. 17.2) is a mobile operating system that was developed by Google and the Open Handset alliance, and it was designed mainly for touchscreen smartphones and tablets. It is built on the Linux kernel, and the first version of the operating system was released in late 2007. The first Android smartphone was released in late 2008, and Android is currently the most widely used operating system.

The source code for Android is released under an open-source licence, and its open-source philosophy has led to a large community of developers who maintain and develop new versions of it. Manufacturers may modify Android as they see fit, and this allows them to customize their devices and differentiate them from competitor products.

Fig. 17.2 Android 6.0

There are over a million applications (apps) for Android, and developers are challenged to ensure that the apps are compatible with the many mobile devices using different hardware and running various (possibly customized) versions of Android.

The iOS operating system is a mobile operating system employed on Apple's mobile devices such as smartphones and tablets. It was created from the MAC OS/X operating system and introduced in 2007. Multitasking for iOS was introduced in 2010 with the release of iOS version 4.0.

17.10 Review Questions

1. What is an operating system?
2. What are the main functions of an operating system?
3. Explain the following operating system concepts: processor scheduling, multiprogramming, paging/segmentation and multitasking.
4. Describe IBM's contributions to operating system development.
5. Describe the similarities and differences between VM and MVS.
6. Describe the influence of the UNIX operating system.
7. Describe the features of DEC's VMS operating system.

17.11 Summary

An operating system is a collection of software programs that control the hardware of a computer and makes it usable. It makes the computing power of the hardware available to the users of the computer, and it manages the hardware to achieve good system performance.

The earliest computers did not have an operating system, and the user had exclusive control over a large computer for a specified period of time. The earliest operating systems were designed in the 1950s with the goal to make more efficient use of the computer (as computers were expensive). These batch-processing systems ran one job at a time, and programs and data were submitted in groups (or batches).

These evolved during the early 1960s into batch multiprogramming systems that were designed to get better utilization of the expensive computer resources. They could handle several diverse jobs at once. However, software development in this environment was very slow. This led operating system designers to develop the concept of multiprogramming in which several jobs are in main memory at once.

IBM announced the System/360 family of computers in 1964, and the computers in the family were designed to use the IBM System/360 operating system (OS/360). OS/360 was a batch-oriented operating system.

UNIX was developed at Bell Labs in the early 1970s. It is a multitasking and multi-user operating system. The IBM PC was introduced in 1981, and IBM outsourced the development of the operating system to a small company called Microsoft. Microsoft had the right to license its operating system, MS/DOS, on IBM compatibles, with PC/DOS (or simply DOS) reserved for IBM personal computers only.

The Macintosh was a paradigm shift for the computer industry when it was introduced in 1984. Its MAC operating system was GUI based, friendly, intuitive and easy to use.

Microsoft Windows is a family of graphical operating systems developed by Microsoft, and it has evolved to become the dominant operating system on laptops and personal computers. It has failed to make an impact on the smartphone operating system market, which is dominated by Apple's iOS and Google's Android operating systems.

The Android operating system was designed mainly for touchscreen smartphones and tablets, and the iOS operating system is a mobile operating system employed on Apple's mobile devices.

History of Software Engineering 18

Abstract
This chapter presents a short history of software engineering from its birth at the Garmisch conference in Germany. The IEEE definition of software engineering is discussed, and it is emphasized that software engineering is a lot more than just programming. We discuss the key challenges in software engineering, as well a number of the high-profile software failures. The waterfall and spiral life cycles are discussed, as well a brief discussion on the Rational Unified Process and the popular Agile methodology. We discuss the key activities in the waterfall model such as requirements, design, implementation, unit, system and acceptance testing.

Key Topics
Standish Chaos Report
Software life cycles
Waterfall model
Spiral model
Rational Unified Process
Agile development
Software inspections
Software testing
Project management
CMMI

G. O'Regan, *Introduction to the History of Computing*, Undergraduate Topics in Computer Science, DOI 10.1007/978-3-319-33138-6_18

18.1 Introduction

The approach to software development in the 1950s and 1960s has been described as the *Mongolian Hordes Approach* by Ince and Andrews [1nA:91]. The 'method' or lack of method was characterized by:

The completed code will always be full of defects.
The coding should be finished quickly to correct these defects.
Design as you code approach.

This philosophy accepted defeat in software development and suggested that irrespective of a solid engineering approach, the completed software would always contain lots of defects and that it therefore made sense to code as quickly as possible and to then identify the defects that would be present, so as to correct them as soon as possible.

It was clear in the late 1960s that the existing approaches to software development were deeply flawed and that there was an urgent need for change. The NATO Science Committee organized two famous conferences to discuss critical issues in software development [Bux:75], with the first conference held at Garmisch, Germany, in 1968, and it was followed by a second conference in Rome in 1969.

Over 50 people from 11 countries attended the Garmisch conference, including Edsger Dijkstra, who did important theoretical work on formal specification and verification. The NATO conferences highlighted problems that existed in the software sector in the late 1960s, and the term *software crisis* was coined to refer to these problems. These included budget and schedule overruns, as well as problems with the quality and reliability of the delivered software.

The conference led to the birth of *software engineering* as a discipline in its own right and the realization that programming is quite distinct from science and mathematics. Programmers are like engineers in that they build software products, and they therefore need education in traditional engineering as well as the latest technologies. The education of a classical engineer includes product design and mathematics. However, often computer science education places an emphasis on the latest technologies rather than the important engineering foundations of designing and building high-quality products that are safe for the public to use.

Programmers therefore need to learn the key engineering skills to enable them to build products that are safe for the public to use. This includes a solid foundation on design and the mathematics required for building safe software products. Mathematics plays a key role in engineering and may assist software engineers in the delivery of high-quality software products. Several mathematical approaches to assist software engineers are described in [ORg:06].

There are parallels between the software crisis in the late 1960s and serious problems with bridge construction in the nineteenth century. Several bridges collapsed or were delivered late or over-budget due to the fact that people involved in their design and construction did not have the required engineering knowledge. This led to bridges that were inadequately designed and constructed, leading to their collapse with the loss of life and endangering the lives of the public.

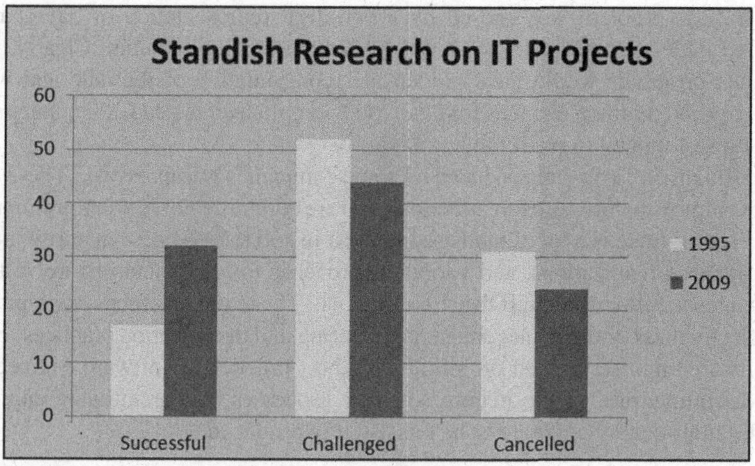

Fig. 18.1 Standish report – results of 1995 and 2009 survey

This led to legislation requiring engineers to be licensed by the Professional Engineering Association prior to practising as engineers. This organization identified a core body of knowledge that the engineer is required to possess, and the licensing body verifies that the engineer has the required qualifications and experience. This helps to ensure that only personnel competent to design and build products actually do so. Engineers have a professional responsibility to ensure that the products are properly built and are safe for the public to use.

The Standish group has conducted research (Fig. 18.1) on the extent of problems with IT projects since the mid-1990s. These studies were conducted in the United States, but there is no reason to believe that European or Asian companies perform any better. The results indicate serious problems with on-time delivery of projects or projects being cancelled prior to completion.[1] However, the comparison between 1995 and 2009 suggests that there have been some improvements with a greater percentage of projects being delivered successfully and a reduction in the percentage of projects being cancelled.

Fred Brooks argues that software is inherently complex and that there is no *silver bullet* that will resolve all of the problems associated with software development such as schedule or budget overruns [Brk:75, Brk:86]. Problems with poor software quality can lead to software flaws that may seriously impact the work of an organization or even loss of life. It is therefore essential that software development organizations place sufficient emphasis on quality throughout the software development life cycle.

[1] These are IT projects covering diverse sectors including banking, telecommunications, etc., rather than pure software companies. Software companies following maturity frameworks such as the CMMI generally achieve more consistent project results, and the CMMI focuses on the management side of software engineering.

The Y2K problem was caused by a two-digit representation of dates, and it required major rework of legacy software for the new millennium. Clearly, well-designed programs would have hidden the representation of the date and would have required minimal changes for year 2000 compliance. Instead, companies spent vast sums of money to rectify the problem.

The quality of software produced by some companies is impressive.[2] These companies employ mature software processes and are committed to continuous improvement. Today, there is a lot of industrial interest in software process maturity models for software organizations, and various approaches to assess and mature software companies are described in [ORg:10, ORg:14].[3] These models focus on improving the effectiveness of the management, engineering and organization practices related to software engineering and on introducing best practice in software engineering. The disciplined use of the mature software processes by the software engineers enables high-quality software to be consistently produced.

18.2 What Is Software Engineering?

Software engineering involves the multi-person construction of multi-version programs. The IEEE 610.12 definition of software engineering is:

> Software engineering is the application of a systematic, disciplined, quantifiable approach to the development, operation, and maintenance of software; that is, the application of engineering to software, and the study of such approaches.

Software engineering includes:

1. Methodologies to design, develop and test software to meet customers' needs.
2. Software is engineered. That is, the software products are properly designed, developed and tested in accordance with engineering principles.
3. Quality and safety are properly addressed.
4. Mathematics may be employed to assist with the design and verification of software products. The level of mathematics employed will depend on the *safety critical* nature of the product. Systematic peer reviews and rigorous testing will often be sufficient to build quality into the software, with heavy *mathematical techniques reserved for safety and security critical software*.
5. Sound project management and quality management practices are employed.
6. Support and maintenance of the software is properly addressed.

[2] I recall projects at Motorola that regularly achieved 5.6σ quality in an L4 CMM environment (i.e. approx. 20 defects per million lines of code. This represents very high quality.).

[3] Approaches such as the CMM or SPICE (ISO 15504) focus mainly on the management and organizational practices required in software engineering. The emphasis is on defining software processes that are fit for purpose and to consistently follow them. The process maturity models focus on what needs to be done rather how it should be done. This gives the organization the freedom to choose the appropriate implementation to meet its needs. The models provide useful information on practices to consider in the implementation.

Software engineering is not just programming. It requires the engineer to state precisely the requirements that the software product is to satisfy and then to produce designs that will meet these requirements. The project needs to be planned and delivered on time and budget. The requirements must provide a precise description of the problem to be solved: i.e. *it should be evident from the requirements what is and what is not required.* The requirements need to be rigorously reviewed to ensure that they are stated clearly and unambiguously and are exactly what the customer wants. The next step is then to create the design that will solve the problem, and it is essential to validate the correctness of the design. Next, the software to implement the design is written, and peer reviews and software testing are employed to verify and validate the correctness of the software.

The verification and validation of the design is rigorously performed for safety critical systems, and it is sometimes appropriate to employ mathematical techniques for this. However, it will often be sufficient to employ peer reviews or software inspections, as these methodologies provide a high degree of rigour. This may include approaches such as Fagan inspections [Fag:76], Gilb inspections [Glb:94] or Prince 2's approach to quality reviews [OGC:04].

The term *engineer* is a title that is awarded on merit in classical engineering. It is generally applied only to people who have attained the necessary education and competence to be called engineers and who base their practice on classical engineering principles. The title places responsibilities on its holder such as to behave professionally and ethically. Often in computer science, the term *software engineer* is employed loosely to refer to anyone who builds things, rather than to an individual with a core set of knowledge, experience and competence.

Several computer scientists (such as Parnas[4]) have argued that computer scientists should be educated as engineers to enable them to apply appropriate scientific principles to their work. They argue that computer scientists should receive a solid foundation in mathematics and design, to enable them to have the professional competence to perform as engineers in building high-quality products that are safe for the public to use. The use of mathematics is an integral part of the engineer's work in other engineering disciplines, and so the *software engineer* should be able to use the appropriate mathematics to assist in the modelling or understanding of the behaviour or properties of a proposed software system.

Software engineers need education[5] on specification, design, turning designs into programs, software inspections and testing. The education should enable the software engineer to produce well-structured programs that are fit for purpose.

---

[4] Parnas has made important contributions to computer science. He advocates a solid engineering approach with the extensive use of classical mathematical techniques to software development. He also introduced information hiding in the 1970s, which is now a part of object-oriented development.

[5] Software companies that are following approaches such as the CMM or ISO 9001 consider the education and qualification of staff prior to assigning staff to performing specific tasks. The appropriate qualifications and experience for the specific role are considered prior to appointing a person to carry out the role. Many companies are committed to the education and continuous development of their staff and on introducing best practice in software engineering into their organization

Parnas has argued that software engineers have responsibilities as professional engineers.[6] They are responsible for designing and implementing high-quality and reliable software that is safe to use. They are also accountable for their decisions and actions[7] and have a responsibility to object to decisions that violate professional standards. Engineers are required to behave professionally and ethically with their clients. The membership of the professional engineering body requires the member to adhere to the code of ethics[8] of the profession. Engineers in other professions are licensed, and therefore Parnas argues that a similar licensing approach be adopted for professional software engineers[9] to provide confidence that they are competent for the particular assignment. Professional software engineers are required to follow best practice in software engineering and the defined software processes.[10]

Many software companies invest heavily in training, as the education and knowledge of its staff are essential to delivering high-quality products and services. Employees receive professional training related to the roles that they are performing, such as project management, service management and software testing. The fact that the employees are professionally qualified increases confidence in the ability of the company to deliver high-quality products and services. A company that pays little attention to the competence and continuous development of its staff will underperform its peers and suffer a loss of reputation and market share.

---

[6] The ancient Babylonians used the concept of accountability, and they employed a code of laws (known as the Hammurabi Code) c. 1750 B.C. It included a law that stated that if a house collapsed and killed the owner, then the builder of the house would be executed.

[7] However, it is unlikely that an individual programmer would be subject to litigation in the case of a flaw in a program causing damage or loss of life. A comprehensive disclaimer of responsibility for problems rather than a guarantee of quality accompanies most software products. Software engineering is a team-based activity involving many engineers in various parts of the project, and it would be potentially difficult for an outside party to prove that the cause of a particular problem is due to the professional negligence of a particular software engineer, as there are many others involved in the process such as reviewers of documentation and code and the various test groups. Companies are more likely to be subject to litigation, as a company is legally responsible for the actions of their employees in the workplace, and a company is a wealthier entity than one of its employees. The legal aspects of licensing software may protect software companies from litigation. However, greater legal protection for the customer can be built into the contract between the supplier and the customer for bespoke-software development.

[8] Many software companies have a defined code of ethics that employees are expected to adhere. Larger companies will wish to project a good corporate image and to be respected worldwide.

[9] The British Computer Scientist (BCS) has introduced a qualification system for computer science professionals that it used to show that professionals are properly qualified. The most important of these is the BCS Information Systems Examination Board (ISEB) which allows IT professionals to be qualified in service management, project management, software testing and so on.

[10] Software companies that are following the CMMI or ISO 9000 standards will employ audits to verify that the processes and procedures have been followed. Auditors report their findings to management and the findings are addressed appropriately by the project team and affected individuals.

18.3 Challenges in Software Engineering

The challenge in software engineering is to deliver high-quality software on time and on budget to customers. The research done by the Standish group was discussed earlier in this chapter, and the results of their 1998 research (Fig. 18.2) on project cost overruns in the United States indicated that 33 % of projects are between 21 % and 50 % overestimate, 18 % are between 51 % and 100 % overestimate and 11 % of projects are between 101 % and 200 % overestimate.

The accurate estimation of project cost, effort and schedule is a challenge in software engineering. Therefore, project managers need to determine how good their estimation process actually is and to make appropriate improvements. The use of software metrics is an objective way to do this, and improvements in estimation will be evident from a reduced variance between estimated and actual effort. The project manager will determine and report the actual versus estimated effort and schedule for the project.

Risk management is an important part of project management, and the objective is to identify potential risks early and throughout the project and to manage them appropriately. The probability of each risk occurring and its impact is determined and the risks are managed during project execution.

Software quality needs to be properly planned to enable the project to deliver a quality product. Flaws with poor quality software lead to a negative perception of the company and may damage the customer relationship and lead to a loss of market share.

There is a strong economic case to building quality into the software, as less time is spent in reworking defective software. The cost of poor quality (COPQ) should be measured and targets set for its reductions. It is important that lessons are learned

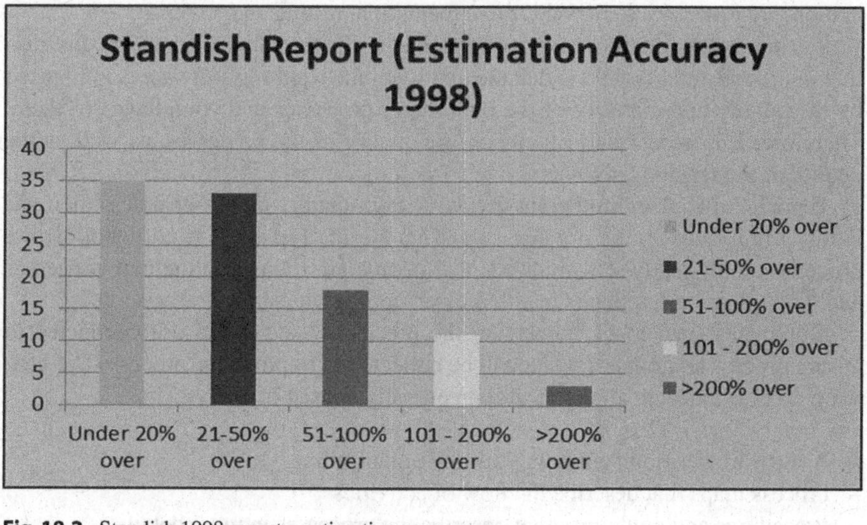

Fig. 18.2 Standish 1998 report – estimation accuracy

during the project and are acted upon appropriately. This helps to promote a culture of continuous improvement.

There have been a number of high-profile software failures [ORg:14]. These included the millennium bug (Y2K) problem, the floating-point bug in the Intel microprocessor, the European Space Agency Ariane-5 disaster and so on. These have caused embarrassment to the organizations as well as the cost of replacement and correction.

The millennium bug was due to the use of two digits to represent dates rather than four digits. The solution involved finding and analysing all code that had a Y2K impact, planning and making the necessary changes and verifying the correctness of the changes. The worldwide cost of correcting the millennium bug is estimated to have been in billions of dollars.

The Intel Corporation was slow to acknowledge the floating-point problem in its Pentium microprocessor and in providing adequate information on its impact to its customers. This led to a large financial cost in replacing microprocessors for its customers. The Ariane-5 failure caused major embarrassment and damage to the credibility of the European Space Agency (ESA). Its maiden flight ended in failure on June 4, 1996, after a flight time of just 40 s.

These failures indicate that quality needs to be carefully considered when designing and developing software. The effect of software failure may be large costs to correct the software, loss of credibility of the company or even loss of life.

18.4 Software Processes and Life Cycles

Organizations vary by size and complexity, and the processes employed will reflect the nature of their business. The development of software involves many processes such as those for defining requirements, processes for project management and estimation and processes for design, implementation, testing and so on.

It is important that the processes employed are fit for purpose, and a key premise in the software quality field is that the quality of the resulting software is influenced by the quality and maturity of the underlying processes and compliance to them. Therefore, it is necessary to focus on the quality of the processes, as well as the quality of the resulting software.

There is, of course, little point in having high-quality processes unless their use is institutionalized in the organization. That is, all employees need to follow the processes consistently. This requires that people are trained on the new processes and that process discipline is instilled by an appropriate audit strategy.

Employees need to be trained on the processes, and audits are conducted to ensure process compliance. Data will be collected to improve the process. The software process assets in an organization generally consist of:

- A software development policy for the organization
- Process maps that describe the flow of activities
- Procedures and guidelines that describe the processes in more detail

- Checklists to assist with the performance of the process
- Templates for the performance of specific activities (e.g. design, testing)
- Training materials

The processes employed to develop high-quality software generally include:

- Project management process
- Requirements process
- Design process
- Coding process
- Peer review process
- Testing process
- Supplier selection processes
- Configuration management process
- Audit process
- Measurement process
- Improvement process
- Customer support and maintenance processes

The software development process has an associated life cycle that consists of various phases. There are several well-known life cycles employed such as the waterfall model [Roy:70], the spiral model [Boe:88], the Rational Unified Process [Jac:99] and the Agile methodology [Bec:00] which has become popular in recent years. The choice of a particular software development life cycle is determined from the particular needs of the specific project. The various life cycles are described in more detail in the following sections.

18.4.1 Waterfall Life Cycle

The origins of the waterfall model[11] (Fig. 18.3) are in the manufacturing and construction industry, and Winston Royce defined it formally for software development in 1970 [Roy:70]. It starts with requirements gathering and definition. It is followed by the functional specification, the design and implementation of the software and comprehensive testing. The testing generally includes unit, system and user acceptance testing.

It is employed for projects where the requirements can be identified early in the project life cycle or are known in advance. It is also called the 'V' life cycle model, with the left-hand side of the 'V' detailing requirements, specification, design and coding and the right-hand side detailing unit tests, integration tests, system tests and acceptance testing. Each phase has entry and exit criteria that must be satisfied before the next phase commences. There are several variations to the waterfall model.

[11] We treat the waterfall model as identical to the V model in this text.

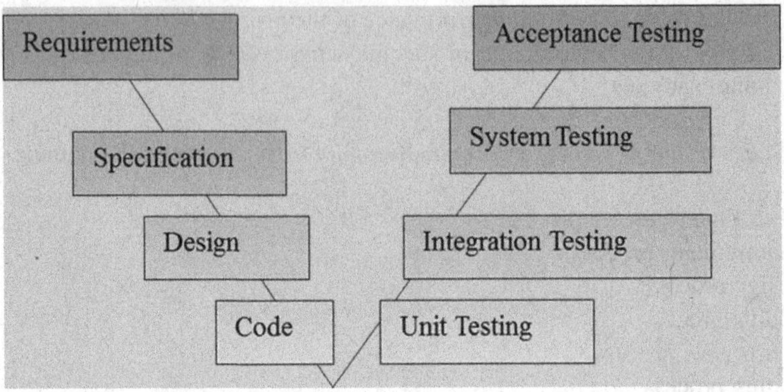

Fig. 18.3 Waterfall V life cycle model

Many companies employ a set of templates to enable the activities in the various phases to be consistently performed. Templates may be employed for project planning and reporting, requirements definition, design, testing and so on. These templates may be based on the IEEE standards or on industrial best practice.

18.4.2 Spiral Life Cycles

The spiral model (Fig. 18.4) was developed by Barry Boehm in the mid-1980s and is useful for a project in which the requirements are not fully known at project initiation, or where the requirements evolve as a part of the development life cycle. The development proceeds in a number of spirals, where each spiral typically involves objectives and an analysis of the risks, updates to the requirements, design, code, testing and a user review of the particular iteration or spiral. The early spirals are concerned with prototyping with the later spirals concerned with the full implementation of the system.

The spiral is, in effect, a reusable prototype with the business analysts and the customer reviewing the current iteration and providing feedback to the development team. The feedback is analysed and used to plan the next iteration. This approach is often used in joint application development, where the usability and look and feel of the application are a key concern. This is important in web-based development and in the development of a graphical user interface (GUI). The implementation of part of the system helps in gaining a better understanding of the requirements of the system, and this feeds into subsequent development cycle. The process repeats until the requirements and the software product are fully complete.

There are several variations of the spiral model including Rapid application development (RAD), joint application development (JAD) models and the dynamic systems development method (DSDM) model. Agile methods have become popular in recent years and these generally employ sprints (or iterations) of 2 weeks duration to implement a number of user stories.

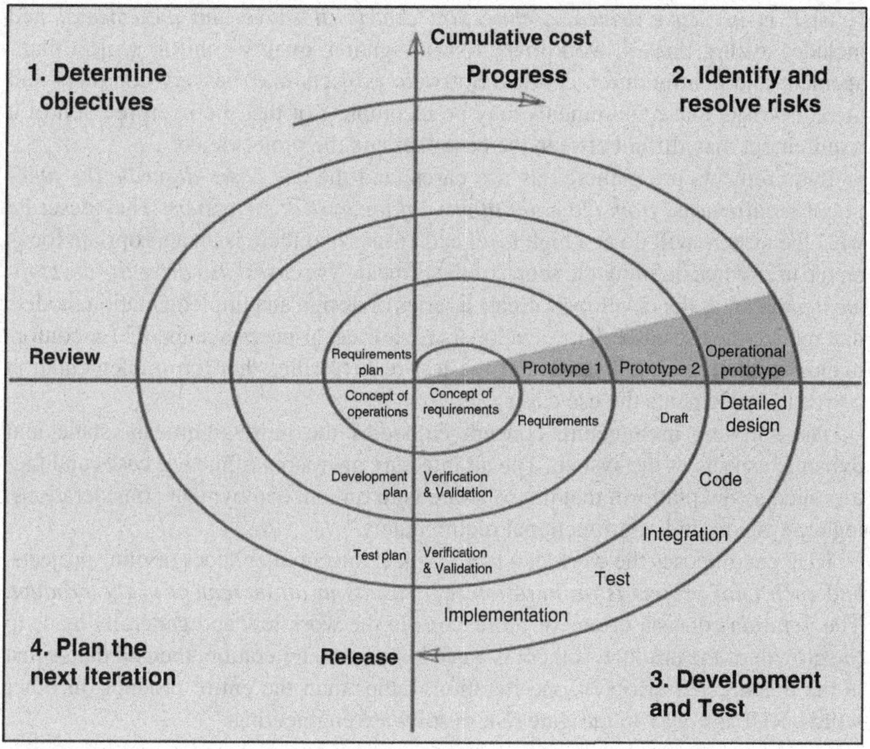

Fig. 18.4 SPIRAL life cycle model. Public domain

There are other life cycle models, for example, the iterative development process that combines the waterfall and spiral life cycle model. The cleanroom approach developed by Harlan Mills at IBM includes a phase for formal specification, and its approach to software testing is based on the predicted usage of the software product. The Rational Unified Process has become popular in recent years, and it is discussed in the next section.

18.4.3 Rational Unified Process

The *Rational Unified Process* [Jac:99] was developed at the Rational Corporation (now part of IBM) in the late 1990s. It uses the Unified Modeling Language (UML) as a tool for specification and design, and UML is a visual modelling language for software systems, which provides a means of specifying, constructing and documenting the object-oriented system. UML was developed by James Rumbaugh, Grady Booch and Ivar Jacobson, and it facilitates the understanding of the architecture and complexity of the system.

RUP is *use case driven*, *architecture centric*, *iterative* and *incremental* and includes cycles, phases, workflows, risk mitigation, quality control, project management and configuration control. Software projects may be very complex, and there are risks that requirements may be incomplete or that the interpretation of a requirement may differ between the customer and the project team.

Requirements are gathered as use cases, and the *use cases describe the functional requirements from the point of view of the user of the system*. They describe what the system will do at a high level and ensure that there is an appropriate focus on the user when defining the scope of the project. *Use cases also drive the development process*, as the developers create a series of design and implementation models that realize the use cases. The developers review each successive model for conformance to the use case model, and the test team verifies that the implementation correctly implements the use cases.

The software architecture concept embodies the most significant static and dynamic aspects of the system. The architecture grows out of the use cases and factors such as the platform that the software is to run on, deployment considerations, legacy systems and non-functional requirements.

RUP decomposes the work of a large project into smaller slices or mini-projects, and *each mini-project is an iteration that results in an increment to the product*. The iteration consists of one or more steps in the workflow and generally leads to the growth of the product. If there is a need to repeat an iteration, then all that is lost is the misdirected effort of one iteration, rather than the entire product. In other words, RUP is a way to mitigate risk in software engineering.

18.4.4 Agile Development

There has been a growth of popularity among software developers in lightweight methodologies such as *Agile*. This is a *software development* methodology that claims to be more responsive to customer needs than traditional methods such as the waterfall model. *The waterfall development model is similar to a wide and slow-moving value stream* and halfway through the project 100 % if the requirements are typically 50 % done. *However, for Agile development 50 % of requirements are typically 100 % done halfway through the project.*

An early version of the methodology was originally introduced in the late 1980s/early 1990s, and the Agile Manifesto was introduced in early 2001 [Bec:01]. Agile has a strong collaborative style of working and its approach includes:

- Aim is to achieve a narrow fast-flowing value stream.
- Feedback and adaptation employed in decision making.
- User stories and sprints are employed.
- Stories are either done or not done.
- Iterative and incremental development is employed.
- A project is divided into iterations.
- An iteration has a fixed length (i.e. time boxing is employed).

- Entire software development life cycle is employed for the implementation of each story.
- Change is accepted as a normal part of life in the Agile world.
- Delivery is made as early as possible.
- Maintenance is seen as part of the development process.
- Refactoring and evolutionary design are employed.
- Continuous integration is employed.
- Short cycle times.
- Emphasis on quality.
- Stand-up meetings.
- Plan regularly.
- Direct interaction is preferred over documentation.
- Rapid conversion of requirements into working functionality.
- Demonstrate value early.
- Early decision-making.

Ongoing changes to requirements are considered normal in the Agile world, and it is believed to be more realistic to change requirements regularly throughout the project rather than attempting to define all of the requirements at the start of the project. The methodology includes controls to manage changes to the requirements, and good communication and early regular feedback are an essential part of the process.

A story may be a new feature or a modification to an existing feature. It is reduced to the minimum scope that can deliver business value, and a feature may give rise to several stories. Stories often build upon other stories and the entire software development life cycle is employed for the implementation of each story. *Stories are either done or not done*: i.e. *there is such thing as a story being 80 % done*. The story is complete only when it passes its acceptance tests. Stories are prioritized based on a number of factors including:

- Business value of story
- Mitigation of risk
- Dependencies on other stories

Sprint planning is performed before the start of the iteration, and stories are assigned to the iteration to fill the available time. The estimates for each story and their priority are determined, and the prioritized stories are assigned to the iteration. *A short morning stand-up meeting is held daily* during the iteration and attended by the project manager and the project team. It discusses the progress made the previous day, problem reporting and tracking and the work planned for the day ahead. A separate meeting is held for issues that require more detailed discussion.

Once the iteration is complete, the latest product increment is demonstrated to an audience including the product owner. This is to receive feedback and to identify new requirements. The team also conducts a retrospective meeting to identify what went well and what went poorly during the iteration. This is for continuous improvement for future iterations.

Agile employs pair programming and a collaborative style of working with the philosophy that two heads are better than one. This allows multiple perspectives in decision-making and a broader understanding of the issues.

Software testing is very important and Agile generally employs automated testing for unit, acceptance, performance and integration testing. Tests are run frequently with the goal of catching programming errors early. They are generally run on a separate build server to ensure that all dependencies are checked. Tests are rerun before making a release. *Agile employs test-driven development with tests written before the code*. The developers write code to make a test pass with ideal developers only coding against failing tests. This approach forces the developer to write testable code.

Refactoring is employed in Agile as a design and coding practice. The objective is to change how the software is written without changing what it does. Refactoring is a tool for evolutionary design where the design is regularly evaluated, and improvements are implemented as they are identified. The automated test suite is essential in showing that the integrity of the software is maintained following refactoring.

Continuous integration allows the system to be built with every change. Early and regular integration allows early feedback to be provided. It also allows all of the automated tests to be run thereby identifying problems earlier.

18.5 Activities in Waterfall Life Cycle

This section describes the various activities in the waterfall software development life cycle in more detail. The activities discussed include:

- Business requirements definition
- Specification of system requirements
- Design
- Implementation
- Unit testing
- System testing
- UAT testing
- Support and maintenance

18.5.1 Business Requirements Definition

The business requirements specify what the customer wants and define what the software system is required to do (*as distinct from how this is to be done*). The requirements are the foundation for the system, and if they are incorrect, then the implemented system will be incorrect. *Prototyping may be employed* to assist in the definition and validation of the requirements.

The specification of the requirements needs to be unambiguous to ensure that all parties involved in the development of the system share a common understanding of what is to be developed and tested.

Requirements gathering involves meetings with the stakeholders to gather all relevant information for the proposed product. The stakeholders are interviewed, and requirements workshops conducted to elicit the requirements from them. An early working system (prototype) is often used to identify gaps and misunderstandings between developers and users. The prototype may serve as a basis for writing the specification.

The requirements workshops with the stakeholders are used to discuss and prioritize the requirements, as well as identifying and resolving any conflicting requirements. The collected information is consolidated into a coherent set of requirements.

The requirements are validated by the stakeholders to ensure that they are actually those desired and to establish their feasibility. This may involve several reviews of the requirements until all stakeholders are ready to approve the requirements document. Changes to the requirements may occur during the project, and these need to be controlled. It is essential to understand the impacts of a change prior to its approval.

The requirements for a system are generally documented in a natural language such as 'English'. Other notations that may be employed to express the requirements include the visual modelling language UML [Jac:05] and formal specification languages such as VDM or Z.

18.5.2 Specification of System Requirements

The specification of the system requirements of the product is essentially a statement of what the software development organization will provide to meet the business requirements. That is, the detailed business requirements are a statement of what the customer wants, whereas the specification of the system requirements is a statement of what will be delivered by the software development organization.

It is essential that the system requirements are valid with respect to the business requirements, and the stakeholders review them to ensure their validity. Traceability may be employed to show how the business requirements are addressed by the system requirements.

There are two categories of system requirements: namely, functional and non-functional requirements. The *functional requirements* define the functionality that is required of the system, and it may include screenshots, report layouts or the desired functionality specified in natural language, use cases, etc. The *non-functional requirements* will generally include security, reliability, performance and portability requirements, as well as usability and maintainability requirements.

18.5.3 Design

The design of the system consists of engineering activities to describe the architecture or structure of the system, as well as activities to describe the algorithms and functions required to implement the system requirements. It is a creative process concerned with how the system will be implemented, and its activities include architecture design, interface design and data structure design. There are often several possible design solutions for a particular system, and the designer will need to decide on the most appropriate solution.

The design may be specified in various ways such as graphical notations that display the relationships between the components making up the design. The notation may include flow charts, or various UML diagrams such as sequence diagrams, state charts and so on. Program description languages or pseudocode may be employed to define the algorithms and data structures that are the basis for implementation.

Functional design involves starting with a high-level view of the system and refining it into a more detailed design. The system state is centralized and shared between the functions operating on that state.

Object-oriented design has become popular in recent years and is based on the concept of *information hiding* [Par:72]. The system is viewed as a collection of objects rather than functions, with each object managing its own state information. The system state is decentralized and an object is a member of a class. The definition of a class includes attributes and operations on class members, and these may be inherited from super classes. Objects communicate by exchanging messages

It is essential to verify and validate the design with respect to the system requirements, and this will be done by design reviews and traceability of the design to the system requirements.

18.5.4 Implementation

This phase is concerned with implementing the design in the target language and environment (e.g. C++ or Java) and involves writing or generating the actual code. The development team divides up the work to be done, with each programmer responsible for one or more modules. The coding activities include code reviews or walkthroughs to ensure that quality code is produced, and to verify its correctness. The code reviews will verify that the source code adheres to the coding standards, that maintainability issues are addressed and that the code produced is a valid implementation of the software design.

Software reuse has become more important in recent times as it provides a way to speed up the development process. Components or objects that may be reused need to be identified and handled accordingly. The implemented code may use software components that have either being developed internally or purchased off the shelf. Open-source software has become popular in recent years, and it allows software developed by others to be used (*under an open-source licence*) in the development of applications.

The benefits of software reuse include increased productivity and a faster time to market. There are inherent risks with customized-off-the shelf (COTS) software, as the supplier may decide to no longer support the software, or there is no guarantee that software that has worked successfully in one domain will work correctly in a different domain. It is therefore important to consider the risks as well as the benefits of software reuse and open-source software.

18.5.5 Software Testing

Software testing is employed to verify that the requirements have been correctly implemented and that the software is fit for purpose, as well as identifying defects present in the software. There are various types of testing that may be conducted including *unit testing, integration testing, system testing, performance testing and user acceptance testing*. These are described below.

18.5.5.1 Unit Testing

Unit testing is performed by the programmer on the completed unit (or module), prior to its integration with other modules. The programmer writes these tests, and the objective is to show that the code satisfies the design. Each unit test case is documented and it should include a test objective and the expected result.

Code coverage and branch coverage metrics are often recorded to give an indication of how comprehensive the unit testing has been. These metrics provide visibility into the number of lines of code executed as well as the branches covered during unit testing.

The developer executes the unit tests, records the results, corrects any identified defects and retests the software. *Test-driven development* has become popular in recent years (e.g. in the Agile world), and this involves writing the unit test case before the code, and the code is written to pass the unit test cases.

18.5.5.2 Integration Test

The development team performs this type of testing on the integrated system, once all of the individual units work correctly in isolation. The objective is to verify that all of the modules and their interfaces work correctly together and to identify and resolve any issues. Modules that work correctly in isolation may fail when integrated with other modules.

18.5.5.3 System Test

The purpose of system testing is to verify that the implementation is valid with respect to the system requirements. It involves the specification of system test cases, and the execution of the test cases will verify that the system requirements have been correctly implemented. An independent test group generally conducts this type of testing, and the system tests are traceable to the system requirements.

Any system requirements that have been incorrectly implemented will be identified, and defects are logged and reported to the developers. The test group will verify that the new version of the software is correct, and regression testing is conducted to verify system integrity. System testing may include security testing, usability testing and performance testing.

The preparation of the test environment requires detailed planning, and it may involve ordering special hardware and tools. It is important that the test environment is set up as early as possible to allow the timely execution of the test cases.

18.5.5.4 Performance Test

The purpose of performance testing is to ensure that the performance of the system is within the bounds specified in the non-functional requirements and to determine if the system is scalable to support future growth. It may include *load performance testing*, where the system is subjected to heavy loads over a long period of time, and *stress testing*, where the system is subjected to heavy loads during a short time interval.

Performance testing often involves the simulation of many users using the system and measuring the response times for various activities. Test tools are employed to simulate a large number of users and heavy loads.

18.5.5.5 User Acceptance Test

UAT testing is usually performed under controlled conditions at the customer site, and its operation will closely resemble the real-life behaviour of the system. The customer will see the product in operation and is able to judge whether or not the system is fit for purpose.

The objective is to demonstrate that the product satisfies the business requirements and meets the customer expectations. Upon its successful completion, the customer is happy to accept the product.

18.5.6 Maintenance

This phase continues after the release of the software product to the customer. Any problems that the customer notes with the software are reported as per the customer support and maintenance agreement. The support issues will require investigation, and the issue may be *a defect in the software*, *an enhancement to the software* or *due to a misunderstanding*. The support and maintenance team will identify the causes of any identified defects and will implement an appropriate solution. Testing is conducted to verify that the solution is correct and that the changes made have not adversely affected other parts of the system. Mature organizations will conduct postmortems to learn lessons from the defect[12] and will take corrective action to prevent a reoccurrence.

The presence of a maintenance phase suggests an acceptance of the reality that problems with the software will be identified post-release. The goal of building a correct and reliable software product the first time is very difficult to achieve, and

[12] This is essential for serious defects that have caused significant inconvenience to customers (e.g. a major telecom outage). The software development organization will wish to learn lessons to determine what went wrong in its processes that prevented the defect from being identified during peer reviews and testing. Actions to prevent a reoccurrence will be identified and implemented.

the customer is always likely to find some issues with the released software product. It is accepted today that quality needs to be built into each step in the development process, with the role of software inspections and testing to identify as many defects as possible prior to release and minimize the risk that that serious defects will be found post-release.

The more effective the in-phase inspections of deliverables, the higher the quality of the resulting implementation, with a corresponding reduction in the number of defects detected by the test groups. The testing group plays a key role in verifying that the system is correct, and in providing confidence that the software is fit for purpose. The approach to software correctness almost seems to be a *brute force* approach, where testing and retesting achieve quality, until the testing group is confident that all defects have been eliminated. Dijkstra [Dij:72] noted that:

> Testing a program demonstrates that it contains errors, never that it is correct.

That is, irrespective of the amount of time spent testing, it can never be said with absolute confidence that the program is correct, and, at best, statistical techniques may be employed to give a measure of the confidence in its correctness. That is, there is no guarantee that all defects have been found in the software.

Many software companies may consider one defect per thousand lines of code (KLOC) to be reasonable quality. However, if the system contains one million lines of code, this is equivalent to a thousand post-release defects, which is unacceptable.

Some mature organizations have a quality objective of three defects per million lines of code. This goal is known as six-sigma (6σ), and Motorola developed it initially for its manufacturing businesses and later applied to its software organization. The goal is to reduce variability in manufacturing processes and to ensure that the processes performed within strict process control limits. Motorola was awarded the first Malcolm Baldrige Quality award for its six-sigma initiative and its commitment to quality.

18.6 Software Inspections

Software inspections are used to build quality into software products, and there are several well-known approaches such as the Fagan methodology [Fag:76], Gilb's approach [Glb:94] and Prince 2's approach.

Fagan inspections were developed by Michael Fagan of IBM. It is a seven-step process that identifies and removes errors in work products. The process mandates that requirements documents, design documents, source code and test plans are all formally inspected by experts independent of the author of the deliverable to ensure quality.

There are various *roles* defined in the process including the *moderator* who chairs the inspection. The *reader's* responsibility is to read or paraphrase the particular deliverable, and *the author* is the creator of the deliverable and has a special interest in ensuring that it is correct. The *tester* role is concerned with the test viewpoint.

The inspection process will consider whether the design is correct with respect to the requirements and whether the source code is correct with respect to the design. Software inspections play an important role in building quality into the software and in reducing the cost of poor quality in the organization. For more detailed information, see [ORg:14].

18.7 Software Project Management

The timely delivery of quality software requires good management and engineering processes. Software projects have a history of being delivered late or over budget, and good project management practices include the following activities:

- Estimation of cost, effort and schedule for the project
- Identifying and managing risks
- Preparing the project plan
- Preparing the initial project schedule and key milestones
- Obtaining approval for the project plan and schedule
- Staffing the project
- Monitoring progress, budget, schedule, effort, risks, issues, change requests and quality
- Taking corrective action
- Replanning and rescheduling
- Communicating progress to affected stakeholders
- Preparing status reports and presentations

The project plan will contain or reference several other plans such as the project quality plan, the communication plan, the configuration management plan and the test plan.

Project estimation and scheduling are difficult as often software projects are breaking new ground and differ from previous projects. That is, previous estimates may often not be a good basis for estimation for the current project. Often, unanticipated problems can arise for technically advanced projects, and the estimates may be optimistic. Gantt charts are generally employed for project scheduling, and these show the work breakdown for the project, as well as task dependencies and the allocation of staff to the various tasks.

The effective management of risk during a project is essential to project success. Risks arise due to uncertainty and the risk management cycle involves[13] risk identification, risk analysis and evaluation, identifying responses to risks, selecting and planning a response to the risk and risk monitoring. The risks are logged, and the likelihood of each risk arising and its impact is then determined. The risk is assigned an owner and an appropriate response to the risk determined. For more detailed information on project management, see [ORg:14].

[13] These are the risk management activities in the Prince 2 methodology.

18.8 CMMI Maturity Model

The CMMI is a framework to assist an organization in the implementation of best practice in software and systems engineering [CKS:11]. It is an internationally recognized model for process improvement and assessment and is used worldwide by thousands of organizations. It provides a framework for an organization to introduce a solid engineering approach to the development of software, and it helps in the definition of high-quality processes for the various software engineering and management activities.

It was developed by the Software Engineering Institute (SEI) who adapted the process improvement principles used in the manufacturing field to the software field. They developed the original CMM model in the early 1990s and its successor the CMMI. The CMMI states *what the organization needs to do* to mature its processes rather than *how this should be done*.

The CMMI consists of five maturity levels with each maturity level consisting of several process areas. Each process area consists of a set of goals, and these goals are implemented by practices related to that process area. Level two is focused on management practices; level three is focused on engineering and organization practices; level four is concerned with ensuring that key processes are performing within strict quantitative limits; level five is concerned with continuous process improvement. Maturity levels may not be skipped in the staged implementation of the CMMI, as each maturity level is the foundation for the next level.

The CMMI allows organizations to benchmark themselves against other organizations. This is done by a formal appraisal conducted by an authorized lead appraiser [SCA:06]. The results of the appraisal are generally reported back to the SEI, and there is a strict qualification process to become an *authorized lead appraiser*. An appraisal is useful in verifying that an organization has improved, and it enables the organization to prioritize improvements for the next improvement cycle. The CMMI is discussed in more detail in [ORg;14].

18.9 Formal Methods

Dijkstra and Hoare have argued that the appropriate way to develop correct software is to derive the program from its formal mathematical specification and to employ *mathematical proof* to demonstrate the correctness of the software with respect to the specification. This offers a rigorous framework to develop programs adhering to the highest-quality constraints. However, in practice mathematical techniques have proved to be cumbersome to use, and their widespread deployment in industry is unlikely at this time.

The *safety-critical area* is one domain to which mathematical techniques have been successfully applied: for example, demonstrating the presence or absence of safety critical properties such as *when a train is in a level crossing, then the gate is closed*. There is a need for extra rigour in the software development process used in the safety critical field, and mathematical techniques can demonstrate the presence or absence of certain desirable or undesirable properties.

Spivey [Spi:92] defines a *formal specification* as the use of mathematical notation to describe in a precise way the properties which an information system must have, without unduly constraining the way in which these properties are achieved. It describes *what* the system must do, as distinct from *how* it is to be done. This abstraction away from implementation enables questions about what the system does to be answered, independently of the detailed code. Furthermore the unambiguous nature of mathematical notation avoids the problem of speculation about the meaning of phrases in an imprecisely worded natural language description of a system.

The formal specification thus becomes the key reference point for the different parties concerned with the construction of the system and is a useful way of promoting a common understanding for all those concerned with the system.

The term *formal methods* is used to describe a formal specification language and a method for the design and implementation of computer systems. The specification is written in a mathematical language, and its precision helps to avoid the problem of ambiguity inherent in a natural language specification. The derivation of an implementation from the specification may be achieved via *step-wise refinement*. Each refinement step makes the specification more concrete and closer to the actual implementation. There is an associated *proof obligation* that the refinement be valid and that the concrete state preserves the properties of the more abstract state. Thus, assuming the original specification is correct and the proofs of correctness of each refinement step are valid, then there is a very high degree of confidence in the correctness of the implemented software.

Formal methods have been applied to a diverse range of applications, including circuit design, artificial intelligence, specification of standards, specification and verification of programs, etc. They are described in more detail in [ORg:06].

18.10 Review Questions

1. Discuss the research results of the Standish group on the current state of IT project delivery.
2. What are the main challenges in software engineering?
3. Describe various software life cycles such as the waterfall model and the spiral model.
4. Discuss the benefits of Agile over conventional approaches. What are the advantages and disadvantages?
5. Describe the purpose of software inspections. What are the benefits?
6. Describe the main activities in software testing.
7. Describe the advantages and disadvantages of formal methods.
8. Describe the main activities in project management.
9. Explain the significance of the CMMI as a framework to improve the software engineering capability of an organization.

18.11 Summary

The birth of software engineering was at the NATO conference held in 1968 in Germany. This conference highlighted the problems that existed in the software sector in the late 1960s, and the term *software crisis* was coined to refer to these. This led to the realization that programming is quite distinct from science and mathematics and that software engineers need to be properly trained to enable them to build high-quality products that are safe to use.

The Standish group conducts research on the extent of problems with the delivery of projects on time and budget. Their research indicates that it remains a challenge to deliver projects on time, on budget and with the right quality.

Programmers are like engineers in the sense that they build products. Therefore, programmers need to receive an appropriate education in engineering as part of their education. Classical engineers receive training on product design and an appropriate level of mathematics.

Software engineering involves multi-person construction of multi-version programs. It is a systematic approach to the development and maintenance of the software, and it requires a precise statement of the requirements of the software product and then the design and development of a solution to meet these requirements. It includes methodologies to design, develop, implement and test software as well as sound project management, quality management and configuration management practices. Support and maintenance of the software is properly addressed.

Software process maturity models such as the CMMI place an emphasis on understanding and improving the software processes in an organization. It is a principle in the software quality field that high-quality processes play a key role in delivering a high-quality product, and the CMMI is a framework that allows high-quality processes to be successfully introduced in the organization. The CMMI allows organizations to benchmark themselves against other similar organizations, and this is done by a formal SCAMPI appraisal conducted by qualified assessors.

Formal methods involve the use of mathematical techniques to provide extra confidence in the correctness of the software. They are used mainly in the safety and security critical fields.

History of Artificial Intelligence

19

Abstract

This chapter presents a short history of artificial intelligence, and we discuss the Turing Test, which is a test of machine intelligence. We discuss strong and weak AI, where strong AI considers an AI programmed computer to be essentially a mind, whereas weak AI considers a computer to simulate thought without real understanding. We discuss Searle's Chinese room, which is a rebuttal of strong AI, and we discuss philosophical issues in AI and Weizenbaum's views on the ethics of AI. There are many subfields in AI and we discuss logic, neural networks and expert systems.

Key Topics

Turing Test
Searle's Chinese room
Philosophical problems in AI
Cognitive psychology
Linguistics
Logic and AI
Robots
Cybernetics
Neural networks
Expert systems

G. O'Regan, *Introduction to the History of Computing*, Undergraduate Topics in Computer Science, DOI 10.1007/978-3-319-33138-6_19

19.1 Introduction

The long-term[1] goal of artificial intelligence is to create a thinking machine that is intelligent, has consciousness, has the ability to learn, has free will and is ethical. The field involves several disciplines such as philosophy, psychology, linguistics, machine vision, cognitive science, mathematics, logic and ethics. Artificial intelligence is a young field and John McCarthy and others coined the term in 1956. Alan Turing had earlier devised the Turing Test as a way to test the intelligent behaviour of a machine. There are deep philosophical problems in artificial intelligence, and some researchers believe that its goals are impossible or incoherent. Hubert Dreyfus and John Searle share these views. Even if artificial intelligence is possible, there are moral issues to consider such as the exploitation of artificial machines by humans and whether it is ethical to do this. Weizenbaum[2] has argued that artificial intelligence is unethical.

One of the earliest references to creating life by artificial means is that of the classical myth of Pygmalion. Pygmalion was a sculptor who carved a woman out of ivory. The sculpture was so realistic that he fell in love with it and offered the statue presents and prayed to Aphrodite the goddess of love. Aphrodite took pity on him and brought the statue (Galathea) to life.

There are several stories of attempts by man to create life from inanimate objects: for example, the creation of the monster in Mary Shelly's Frankenstein. The monster is created by an over ambitious scientist who is punished for his blasphemy of creation (in that creation is for God alone). The monster feels rejected following creation and inflicts a horrible revenge on its creator.

The Czech play 'Rossum's Universal Robots' is a science fiction play by Capek, and it was performed in Prague in 1921. It was translated into English and appeared in London in 1923. It contains the first reference to the term *robot*, and the play considers the exploitation of artificial workers in a factory. The robots (or androids) are initially happy to serve humans, but become unhappy with their existence over a period of time. The fundamental question that the play is considering is whether the robots are being exploited and, if so, whether this is ethical and what should the response of the robots be to their exploitation. It eventually leads to a revolt by the robots and the extermination of the human race.

19.2 Descartes

René Descartes (Fig. 19.1) was an influential French mathematician, scientist and philosopher. He was born in a village in the Loire valley in France in 1596 and studied law at the University of Poitiers. He never practised as a lawyer and instead

[1] This long-term goal may be hundreds of years as there is unlikely to be an early breakthrough in machine intelligence as there are deep philosophical problems to be solved.

[2] Weizenbaum was a psychologist who invented the ELIZA program, which simulated a psychologist in dialogue with a patient. He was initially an advocate of artificial intelligence but later became a critic.

Fig. 19.1 Rene Descartes

served Prince Maurice of Nassau in the Netherlands. He invented the Cartesian coordinate system that is used in plane geometry and algebra. In this system, each point on the plane is identified through a pair of numbers (x, y): the x-coordinate and the y-coordinate.

He made important contributions to philosophy and attempted to derive a fundamental set of principles that can be known to be true. His approach was to renounce any idea that could be doubted. He rejected the senses since they can deceive and are not a sound source of knowledge. For example, during a dream the subject perceives stimuli that appear to be real, but these have no existence outside the subject's mind. Therefore, it is inappropriate to rely on one's senses as the foundation of knowledge.

He argued that a powerful *evil demon or mad scientist* could exist who sets out to manipulate and deceive subjects, thereby preventing them from knowing the true nature of reality. The evil demon could bring the subject into existence including an implanted memory. The question is how one can know for certain what is true given the limitations of the senses. The *brain in the vat thought experiment* is a more modern formulation of the idea of an evil spirit or mad scientist. A mad scientist could remove a person's brain from their body and place it in a vat and connects its neurons by wires to a supercomputer. The computer provides the disembodied brain with the electrical impulses that the brain would normally receive. The computer could then simulate reality, and the disembodied brain would have conscious experiences and would receive the same impulses as if it were inside a person's skull. There is no way to tell whether the brain is inside the vat or inside a person.

That is, at any moment an individual could potentially be a brain connected to a sophisticated computer program or inside a person's skull. Therefore, if you cannot

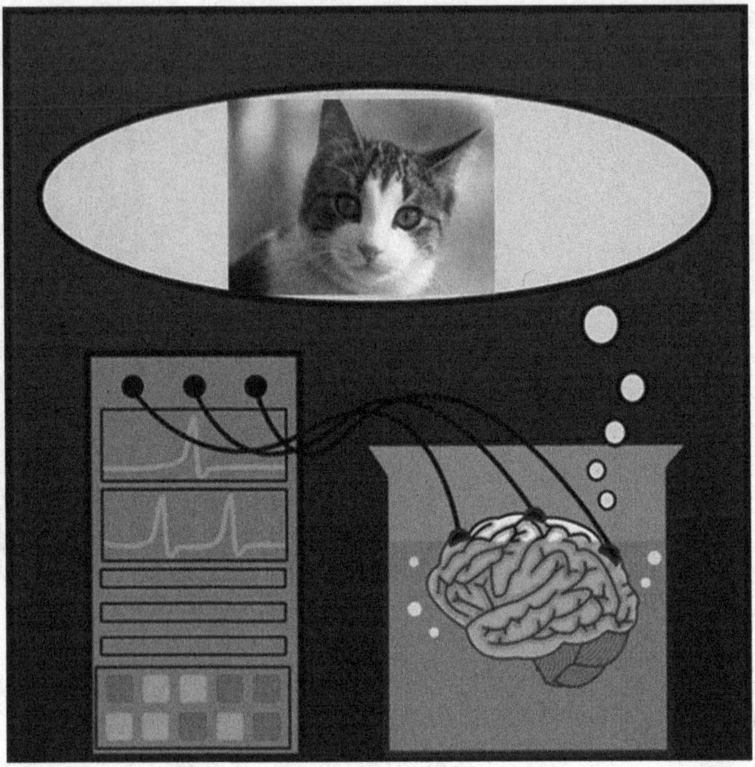

Fig. 19.2 Brain in a VAT thought experiment

be sure that you are not a brain in a vat, then you cannot rule out the possibility that all of your beliefs about the external world are false. This sceptical argument is difficult to refute.

The perception of a 'cat' (Fig. 19.2) in the case where the brain is in the vat is false and does not correspond to reality. It is impossible to know whether your brain is in a vat or inside your skull; it is therefore impossible to know whether your belief is valid or not.

From this, Descartes deduced that there was one single principle that must be true. He argued that even if he is being deceived, then clearly he is thinking and must exist. This principle of existence or being is more famously known as '*cogito, ergo sum*' (I think, therefore I am). Descartes argued that this existence can be applied to the present only, as memory may be manipulated and therefore doubted. Further, the only existence that he is sure of is that he is a *thinking thing*. He cannot be sure of the existence of his body, as his body is perceived by his senses, which he has proven to be unreliable. Therefore, his mind or thinking thing is the only thing about him that cannot be doubted. His mind is used to make judgements and to deal with unreliable perceptions received via the senses.

Descartes constructed a system of knowledge (*rationalism*) from this one principle using the deductive method. He deduced the existence of a benevolent God using the ontological argument. He argues [Des:99] that we have an innate idea of a supremely perfect being (God) and that God's existence may be inferred immediately from the innate idea of a supremely perfect being.

1. I have an innate idea of a supremely perfect being (i.e. God).
2. Necessarily, existence is a perfection.[3]
3. Therefore God exists.

He then argued that since God is benevolent, he can have some trust in the reality that his senses provide. God has provided him with a thinking mind and does not wish to deceive him. He argued that knowledge of the external world can be obtained by both perception and deduction and that reason (or rationalism) is the only reliable method of obtaining knowledge.

Descartes was a *dualist* and he makes a clear *mind-body* distinction. He states that there are two substances in the universe: mental substances and bodily substances. The *mind-body distinction is very relevant in AI* and the analogy of the human mind and brain is software running on a computer.

This thinking thing (*res cogitans* or mind/soul) is distinct from the rest of nature (*res extensa*) and interacts with the world through the senses to gain knowledge. Knowledge is gained by mental operations using the deductive method, where starting from the premises that are known to be true, further truths may be logically deduced. Descartes founded what would become known as the rationalist school of philosophy where knowledge was derived solely by human reasoning. The *analogy of the mind in AI would be an AI program running on a computer* with knowledge gained by sense perception with sensors and logical deduction.

Descartes believed that the bodies of animals are complex living machines without feelings. He dissected (including vivisection) many animals for experiments. His experiments led him to believe that the actions and behaviour of non-human animals can be fully accounted for by mechanistic means, without reference to the operations of the mind. He realized from his experiments that a lot of human behaviour (e.g. physiological functions and blinking) is like that of animals in that it has a mechanistic explanation.

Descartes was of the view that well-designed automata[4] could mimic many parts of human behaviour. He argued that the key differentiators between human and animal behaviour were that humans could adapt to widely varying situations and also had the ability to use language. The use of language illustrates the power of the use of thought, and it clearly differentiates humans from animals. Animals do not possess the ability to use language for communication or reason. This, he argues,

[3] Descartes' ontological argument is similar to St. Anselm's argument on the existence of God, and implicitly assumes existence as a predicate (which was refuted by Kant).

[4] An automaton is a self-operating machine or mechanism that behaves and responds in a mechanical way.

provides evidence for the presence of a soul associated with the human body. In essence, animals are pure machines, whereas humans are machines with minds (or souls).

The significance of Descartes in the field of artificial intelligence is that the Cartesian dualism that humans seem to possess would need to be reflected among artificial machines. Humans seem to have a distinct sense of 'I' as distinct from the body, and the 'I' seems to represent some core sense or essence of being that is unchanged throughout the person's life. It somehow represents personhood, as distinct from the physical characteristics of a person that are inherited genetically. The long-term challenge for the AI community is to construct a machine that (in a sense) possesses Cartesian dualism: i.e. a machine that has awareness of itself as well as its environment.

19.3 The Field of Artificial Intelligence

The origin of the term 'artificial intelligence' is in work done on the proposal for Dartmouth Summer Research Project on Artificial Intelligence. John McCarthy and others wrote this proposal in 1955, and the research project took place in the summer of 1956.

The success of early AI went to its practitioners' heads and they believed that they would soon develop machines that would emulate human intelligence. They convinced many of the funding agencies and the military to provide research grants, as they believed that real artificial intelligence would soon be achieved. They had some initial (limited) success with machine translation, pattern recognition and automated reasoning. However, it is now clear that AI is a long-term project. Artificial intelligence is a multidisciplinary field and includes disciplines such as:

- Computing
- Logic and philosophy
- Psychology
- Linguistics
- Neuroscience and neural networks
- Machine vision
- Robotics
- Expert systems
- Machine translation
- Epistemology and knowledge representation

The British mathematician, Alan Turing, contributed to the debate concerning thinking machines, consciousness and intelligence in the early 1950s [Tur:50]. He devised the famous 'Turing Test' to judge whether a machine was conscious and intelligent. Turing's paper was very influential as it raised the idea of the possibility of programming a computer to behave intelligently.

Fig. 19.3 John McCarthy (Courtesy of John McCarthy)

Shannon considered the problem of writing a chess program in the late 1940s, and he distinguished between a brute force strategy where the program could look at every combination of moves and a strategy where knowledge of chess could be used to select and examine a subset of available moves. The ability of a program to play chess is a skill that is considered intelligent, even though the machine itself is not conscious that it is playing chess.

Modern chess programs have been quite successful and have advantages over humans in terms of computational speed in considering combinations of moves. The IBM chess program 'Deep Blue' defeated Kasparov in 1997.

Herbert Simon and Alan Newell developed the first theorem prover with their work on a program called 'Logic Theorist' or 'LT' [NeS:56]. This program could independently provide proofs of various theorems in Russell's and Whitehead's *Principia Mathematica*[5] [RuW:10]. LT was demonstrated at the Dartmouth conference, and it showed that computers had the ability to encode knowledge and information and to perform intelligent operations such as solving theorems in mathematics.

John McCarthy (Fig. 19.3) proposed a program called the advice taker in his influential paper 'Programs with Common Sense' [Mc:59]. The idea was that this program would be able to draw conclusions from a set of premises, and McCarthy

[5] Russell is said to have remarked that he was delighted to see that the *Principia Mathematica* could be done by machine and that if he and Whitehead had known this in advance, they would not have wasted 10 years doing this work by hand in the early twentieth century. The LT program succeeded in proving 38 of the 52 theorems in Chap. 2 of *Principia Mathematica*. Its approach was to start with the theorem to be proved and to then search for relevant axioms and operators to prove the theorem.

states that a program has common sense if it is capable of automatically deducing for itself *a sufficiently wide class of immediate consequences of anything it is told and what it already knows*.

The advice taker uses logic to represent knowledge (i.e. premises that are taken to be true), and it then applies the deductive method to deduce further truths from the relevant premises.[6] That is, the program manipulates the formal language (e.g. predicate logic) and provides a conclusion that may be a statement or an imperative. McCarthy envisaged that the advice taker would be a program that would be able to learn and improve. This would involve making statements to the program and telling it about its symbolic environment. The program will have all the logical consequences of what it has already been told and the previous knowledge. McCarthy's desire was to create programs to learn from their experience as effectively as humans do.

The McCarthy philosophy is that common sense[7] knowledge, reasoning and problem solving can be formalized with logic. A particular system is described by a set of sentences in logic. These logic sentences represent all that is known about the world in general and what is known about the particular situation and the goals of the systems. The program then performs actions that it infers are appropriate for achieving its goals.

19.3.1 Turing Test and Strong AI

Alan Turing contributed to the debate concerning artificial intelligence in his 1950 paper on computing, machinery and intelligence [Tur:50]. Turing's paper considered whether it could be possible for a machine to be conscious and to think. He predicted that it would be possible to speak of machines thinking, and he devised a famous experiment that would determine if a computer had these attributes. This is known as the *Turing Test*, and it is an adaptation of a well-known party game, which involves three participants. One of them, the judge, is placed in a separate room from the other two: one is a male and the other is a female. Questions and responses are typed and passed under the door. The objective of the game is for the judge to determine which participant is male and which is female. The male is allowed to deceive the judge, whereas the female is supposed to assist.

Turing adapted this game by allowing the role of the male to be played by a computer. The test involves a judge who is engaged in a natural language conversation with two other parties: one party is a human and the other is a machine. If the judge cannot determine which is machine and which is human, then the machine is said to have passed the 'Turing Test'. That is, a machine that passes the Turing Test must be considered intelligent, as it is indistinguishable from a human. The test is

[6] Of course, the machine would somehow need to know what premises are relevant and should be selected in applying the deductive method from the many premises that are already encoded.

[7] Common sense includes basic facts about events, beliefs, actions, knowledge and desires. It also includes basic facts about objects and their properties.

applied to test the linguistic capability of the machine rather than the audio capability, and the conversation is limited to a text-only channel.

Turing's work on 'thinking machines' led to a debate concerning the nature of intelligence, and it caused a great deal of public controversy as defenders of traditional values attacked the idea of machine intelligence.

Turing strongly believed that machines would eventually be developed that would stand a good chance of passing the 'Turing Test'. He considered the operation of 'thought' to be equivalent to the operation of a discrete state machine. A program that runs on a single, universal machine, i.e. a computer, may simulate such a machine.

Turing viewpoint that a machine will one day pass the Turing Test and be considered intelligent is known as *strong artificial intelligence*. It states that a computer with the right program would have the mental properties of humans. There are a number of objections to strong AI, and one well-known rebuttal is that of Searle's Chinese room argument.

Searle's Chinese room thought experiment is a famous paper on machine understanding [Sea:80]. This classic paper presents a compelling argument against the feasibility of the strong AI project. It rejects the claim that a machine will someday in the future have the same cognitive qualities as humans. Searle argues that *brains cause minds, and that syntax does not suffice for semantics*. He defines the terms *strong* and *weak AI* as follows:

Strong AI

> The computer is not merely a tool in the study of the mind, rather the appropriately programmed computer really *is* a mind in the sense that computers given the right programs can be literally said to *understand* and have other cognitive states. [Searle's 1980 Definition]

Weak AI

> Computers just *simulate* thought, their seeming understanding isn't real understanding (just as-if), their seeming calculation is only as-if calculation, etc. Nevertheless, computer simulation is useful for *studying* the mind (as for studying the weather and other things).

19.3.1.1 The Chinese Room Thought Experiment

A man is placed into a closed room into which Chinese writing symbols are input to him (Fig. 19.4). He is given a rulebook that shows him how to manipulate the symbols to produce Chinese output. He has no idea as to what each symbol means, but with the rulebook, he is able to produce the Chinese output. This allows him to communicate with the other person and appear to understand Chinese. The rulebook allows him to answer any questions posed, without the slightest understanding of what he is doing or what the symbols mean.

1. Chinese characters are entered through slot 1.
2. The rulebook is employed to construct new Chinese characters.
3. Chinese characters are outputted to slot 2.

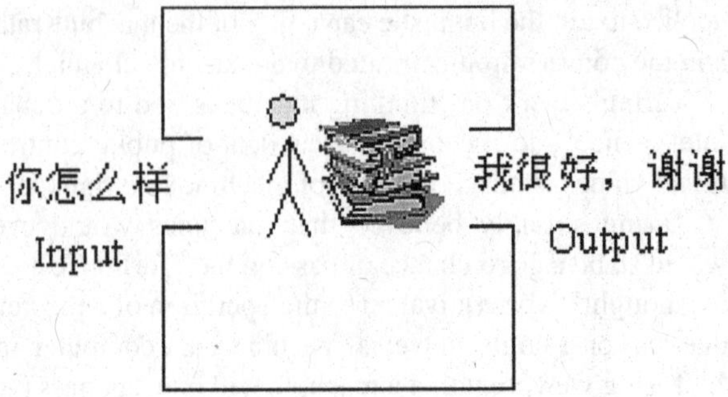

Fig. 19.4 Searle's Chinese room

The question *Do you understand Chinese?* could potentially be asked, and the rulebook would be consulted to produce the answer 'Yes, of course' despite of the fact that the person inside the room has not the faintest idea of what is going on. It will appear to the person outside the room that the person inside is knowledgeable on Chinese. The person inside is just following rules without understanding.

The process is essentially that of a computer program that takes an input, performs a computation based on the input and then finally produces an output. Searle has essentially constructed a machine that can never be mental. Changing the program essentially means changing the rulebook, and this does not increase understanding. The strong artificial intelligence thesis states that given the right program, *any* machine running it would be mental. However, Searle argues that the program for this Chinese room would not understand anything and that therefore the strong AI thesis must be false. In other words, Searle's Chinese room argument is a rebuttal of strong AI by showing that a program running on a machine that appears to be intelligent has no understanding whatsoever of the symbols that it is manipulating. That is, given any rulebook (i.e. program), the person would never understand the meanings of those characters that are manipulated.

That is, just because the machine acts like it knows what is going on, it actually only knows what it is programmed to know. It differs from humans in that it is not aware of the situation like humans are. It suggests that machines may not have intelligence or consciousness, and the Chinese room argument applies to any Turing equivalent computer simulation.

There are several rebuttals of Searle's position and one well-known rebuttal attempt is the 'System Reply' argument. This reply argues that if a result associated with intelligence is produced, then intelligence must be found somewhere in the system. The proponents of this argument draw an analogy between the human brain and its constituents. None of its constituents have intelligence, but the system as a whole (i.e. the brain) exhibits intelligence. Similarly, the parts of the Chinese room may lack intelligence, but the system as a whole is intelligence.

19.3.2 Ethics and AI

Weizenbaum created the famous Eliza program at MIT in 1966, and the program interacted with a user sitting at an electric typewriter, in English, in the manner of a Rogerian psychotherapist.[8]

The name *Eliza*[9] was chosen because it could be taught to speak increasingly well, as in Shaw's Pygmalion. It was one of the earliest natural language processing programs, and Weizenbaum gave an example of a conversation between Eliza and a young woman in [Wei:66]. Many users believed that Eliza had real understanding, and they began to unburden themselves in long computer sessions.

The program operated by decomposing the user's input into its constituent parts of speech and then typing them back at the user in a manner to continue with the dialogue. The program provided very convincing human-like interaction, as may be seen in the sample dialogue below:

User/patient:	Men are all alike.
Eliza:	*In what way* ?
User/patient:	They are always bugging us about something or others.
Eliza:	*Can you think of a specific example*?
User/patient:	Well, my boyfriend made me come here.
Eliza:	*Your boyfriend made you come here*?
User/patient:	He says I'm depressed much of the time.
Eliza:	*I am sorry to hear that you are depressed.*
User/patient:	It's true; I am unhappy.
Eliza:	*Do you think that coming here will help you not to be unhappy*?
User/patient:	I need some help; that much seems certain.
Eliza:	*What would it mean to you if you got some help* ?
User/patient:	Perhaps, I could learn to get along with my mother.
Eliza:	*Tell me more about your family.*

Weizenbaum was shocked to discover that so many users were convinced that the program had real understanding, and users spent hours sharing their personal problems with the program. It led Weizenbaum to think about the ethics and implications of the artificial intelligence field and the ability of a relatively trivial program to deceive a naïve user to reveal personal information. He became a leading critic of the AI field and an advocate for professional and social responsibility.

His views on the ethics of AI are discussed in his book *Computer Power and Human Reason* [Wei:76]. He displays ambivalence towards computer technology, and he argues that AI is a threat to human dignity and that AI should not replace humans in positions that require respect and care. He states that machines lack

[8] Rogerian psychotherapy (person-centred therapy) was developed by Carl Rodgers in the 1940s.

[9] Eliza Doolittle was a working-class character in Shaw's play Pygmalion. She is taught to speak with an upper-class English accent.

empathy and that if they replace humans in positions such as police officers or judges, this would lead to alienation and a devaluation of the human condition.

His ELIZA program demonstrated the threat that AI poses to privacy. It is conceivable that an AI program may be developed in the future that is capable of understanding speech and natural languages. Such a program could theoretically eavesdrop on every phone conversation and email and gather private information on what is said and who is saying it. Such a program could be used by a state to suppress dissent and to eliminate those who pose a threat.

As more and more sophisticated machines and robots are created, it is, of course, essential that intelligent machines behave ethically and have a moral compass to distinguish right from wrong. It remains an open question as to how to teach a robot right from wrong.

19.4 Philosophy and AI

Artificial intelligence includes the study of knowledge and the mind, and there are deep philosophical problems (e.g. the nature of mind, consciousness and knowledge) to be solved.

The Greeks did important early work on philosophy as they attempted to understand the world and the nature of being and reality. Thales and the Milesians[10] attempted to find an underlying principle that would explain the nature of the world. Pythagoras believed that mathematics was this basic principle and that everything (e.g. music) could be explained in terms of number. Plato distinguished between the world of appearances and the world of reality. He argued that the world of appearances resembles the flickering shadows on a cave wall, whereas reality is in the world of ideas[11] or forms, in which objects of this world somehow participate. Aristotle proposed the framework of a substance, which includes form plus matter. For example, the matter of a wooden chair is the wood that it is composed of, and its form is the general form of a chair.

Descartes had a significant influence on the philosophy of mind and AI. Knowledge is gained by mental operations using the deductive method. This involves starting from premises that are known to be true and deriving further truths. He distinguished between the mind and the body (Cartesian dualism), and the analogy of the mind is an AI program running on a computer with sensors and logical deduction used to gain knowledge.

British Empiricism rejected the Rationalist position and stressed the importance of empirical data in gaining knowledge about the world. It argued that all knowledge is derived from sense experience. It included philosophers such as Locke,

[10] The term 'Milesians' refers to inhabitants of the Greek city-state Miletus which is located in modern Turkey. Anaximander and Anaximenes were two other Milesians who made contributions to early Greek philosophy approx 600 B.C.

[11] Plato was an idealist: i.e. that reality is in the world of ideas rather than the external world. Realists (in contrast) believe that the external world corresponds to our mental ideas.

Berkeley[12] and Hume. Locke argued that a child's mind is a blank slate (*tabula rasa*) at birth and that all knowledge is gained by sense experience. Berkeley argued that the ideas in a man's mind have no existence outside his mind [Ber:99], and this philosophical position is known as idealism.[13] David Hume formulated the standard empiricist philosophical position in 'An Enquiry Concerning Human Understanding' [Hum:06].

Hume argued that all objects of human knowledge may be divided into two kinds: *matters of fact* propositions that are based entirely on experience or *relation of ideas* propositions that may be demonstrated via deduction reasoning in the operations of the mind. He argued that any subject[14] proclaiming knowledge that does not adhere to these empiricist principles *should be committed to the flames*[15] *as such knowledge contains nothing but sophistry and illusion.*

Kant's *Critique of Pure Reason* [Kan:03] was published in 1781 and is a response to Hume's theory of empiricism. Kant argued that there is a third force in human knowledge that provides concepts that can't be inferred from experience. Such concepts include the laws of logic (e.g. modus ponens), causality and so on, and Kant argued that the third force was the manner in which the human mind structures its experiences. These structures are called categories.

The continental school of philosophy included thinkers such as Heidegger and Merleau-Ponty who argued that the world and the human body are mutually intertwined. Merleau-Ponty emphasized the concept of a body-subject that actively participates both as the perceiver of knowledge and as an object of perception. Heidegger emphasized that existence can only be considered with respect to a changing world.

---

[12] Berkeley was an Irish philosopher and he was born in Dysart castle in Kilkenny, Ireland. He was educated at Trinity College, Dublin, and served as bishop of Cloyne in Co. Cork. He planned to establish a seminary in Bermuda for the sons of colonists in America, but the project failed due to lack of funding from the British government. Berkeley University in San Francisco is named after him.

[13] Berkeley's theory of ontology is that for an entity to exist, it must be perceived: i.e. '*Esse est percipi*'. He argues that 'It is an opinion strangely prevailing amongst men, that houses, mountains, rivers, and in a world all sensible objects have an existence natural or real, distinct from being perceived'. This led to a famous Limerick that poked fun at Berkeley's theory. 'There once was a man who said God; Must think it exceedingly odd; To find that this tree, continues to be; When there is no one around in the Quad'. The reply to this Limerick was appropriately 'Dear sir, your astonishments odd; I am always around in the Quad; And that's why this tree will continue to be; Since observed by, yours faithfully, God'.

[14] Hume argues that these principles apply to subjects such as theology as its foundations are in faith and divine revelation, which are neither matters of fact nor relations of ideas.

[15] 'When we run over libraries, persuaded of these principles, what havoc must we make? If we take in our hand any volume; of divinity or school metaphysics, for instance; let us ask, *Does it contain any abstract reasoning concerning quantity or number?* No. *Does it contain any experimental reasoning concerning matter of fact and existence?* No. Commit it then to the flames: for it can contain nothing but sophistry and illusion'.

Philosophy has been studied for over two millennia, but to date very little progress has been made in solving its fundamental questions. However, it is important that it be considered as any implementation of AI will make philosophical assumptions and it is important that these be understood.

19.5 Cognitive Psychology

Psychology arose out of the field of psychophysics in the late nineteenth century with the work by German pioneers in attempting to quantify perception and sensation. Fechner's mathematical formulation of the relationship between stimulus and sensation is given by

$$S = k \log I + c$$

The symbol S refers to the intensity of the sensation, the symbols k and c are constants, and the symbol I refers to the physical intensity of the stimulus. William James defined psychology as the science of mental life.

One of the early behaviouralist psychologists was Pavlov who showed that it was possible to develop a conditional reflex in a dog. He showed that it is possible to make a dog salivate in response to the ringing of a bell. This is done by ringing a bell each time before meat is provided to the dog, and the dog therefore associates the presentation of meat with the ringing of the bell after a training period.

Skinner developed the concept of conditioning further using rewards to reinforce desired behaviour and punishment to discourage undesired behaviour. Positive reinforcement helps to motivate the individual to behave in the desired way, with punishment used to deter the individual from performing undesired behaviour. The behavioural theory of psychology explains many behavioural aspects of the world. However, it does not really explain complex learning tasks such as language development.

Merleau-Ponty[16] considered the problem of what the structure of the human mind must be for the objects of the external world to exist in our minds in the form that they do. He built upon the theory of phenomenology as developed by Hegel and Husserl. Phenomenology involves a focus and exploration of phenomena with the goal of establishing the essential features of experience. Merleau-Ponty introduced the concept of the *body-subject*, which is distinct from the Cartesian view that the world is just an extension of our own mind. He argued that the world and the human body are mutually intertwined. The Cartesian view is that the self must first be aware of and know its own existence, prior to being aware of and recognizing the existence of anything else.

The body has the ability to perceive the world, and it plays a double role in that it is both the subject (i.e. the perceiver) and the object (i.e. the entity being perceived)

[16] Merleau-Ponty was a French philosopher who was strongly influenced by the phenomenology of Husserl. He was also closely associated with the French existentialist philosophers such as Jean-Paul Sartre and Simone De Beauvoir.

of experience. Human understanding and perception are dependent on the body's capacity to perceive via the senses and its ability to interrogate its environment. Merleau-Ponty argued that there is a symbiotic relationship between the perceiver and what is being perceived, and he argues that as our consciousness develops, the self imposes richer and richer meanings on objects. He provides a detailed analysis of the flow of information between the body-subject and the world.

Cognitive psychology is a branch of psychology that is concerned with learning, language, memory and internal mental processes. Its roots lie in Piaget's child development psychology and in Wertheimer's Gestalt psychology. The latter argues that the operations of the mind are holistic and that the mind contains a self-organizing mechanism. Holism argues that the sum of the parts is less than the whole, and it is the opposite of logical atomism[17] developed by Bertrand Russell. Russell (and also Wittgenstein) attempted to identify the atoms of thought: i.e. the elements of thought that cannot be divided into smaller pieces. Logical atomism argued that all truths are ultimately dependent on a layer of atomic facts. It had an associated methodology whereby by a process of analysis, it attempted to construct more complex notions in terms of simpler ones.

Cognitive psychology was developed in the late 1950s and is concerned with how people understand, diagnose and solve problems, as well as the mental processes that take place during a stimulus and its corresponding response. It argues that solutions to problems take the form of rules, heuristics and sudden insight, and it considers the mind as having a certain conceptual structure. The dominant *paradigm* in the field has been the *information processing model*, which considers the mental processes of thinking and reasoning as being equivalent to software running on the computer: i.e. the brain. It has associated theories of input, representation of knowledge, processing of knowledge and output.

Cognitive psychology has been applied to artificial intelligence from the 1960s, and some of the research areas include:

- Perception
- Concept formation
- Memory
- Knowledge representation
- Learning
- Language
- Grammar and linguistics
- Thinking
- Logic and problem solving

It is clear that for a machine to behave with intelligence, it will need to be able to perceive objects in the physical world. It must be able to form concepts and to

[17] Atomism actually goes back to the work of the ancient Greeks and was originally developed by Democritus and his teacher Leucippus in the fifth century B.C. Atomism was rejected by Plato in the dialogue the *Timaeus*.

remember knowledge that it has already been provided with. It will need an understanding of temporal events. Knowledge must be efficiently represented to allow easy retrieval for analysis and decision-making. An intelligent machine will need the ability to produce and understand written or spoken language. A thinking machine must be capable of thought, learning, analysis and problem solving.

19.6 Computational Linguistics

Linguistics is the theoretical and applied study of *language, and human language is highly complex. It* includes the study of phonology, morphology, syntax, semantics and pragmatics. Syntax is concerned with the study of the rules of grammar, and the application of the rules forms the syntactically valid sentences and phrases of the language. Morphology is concerned with the formation and alteration of *words*, and phonetics is concerned with the study of sounds and how sounds are produced and perceived as speech (or non-speech).

Noam Chomsky is considered the father of linguistics, and he has been highly influential in the linguistics field. He defined the Chomsky Hierarchy of grammars [ORg:13], which classifies grammars into a number of classes with increasing expressive power. These consist of four levels including regular grammars, context-free grammars, context-sensitive grammars and unrestricted grammars. Each successive class can generate a broader set of formal languages than the previous. The grammars are distinguished by their production rules, which determine the type of language that is generated.

Computational linguistics is an *interdisciplinary* study of the design and analysis of natural language processing systems. It includes linguists, *computer scientists, cognitive psychologists, mathematicians* and experts in *artificial intelligence*.

Early work on computational linguistics commenced with machine translation work in the United States in the 1950s. The objective was to develop an automated mechanism by which Russian language texts could be translated directly into English without human intervention. It was naively believed that it was only a matter of time before automated machine translation would be done.

However, the initial results were not very successful, and it was realized that the automated processing of human languages was considerably more complex. This led to the birth of a new field called computational linguistics, and the objective of this field is to investigate and develop algorithms and software for processing natural languages. It is a subfield of artificial intelligence and deals with the comprehension and production of natural languages.

The task of translating one language into another requires an understanding of the grammar of both languages. This includes an understanding of the syntax, the morphology, semantics and pragmatics of the language. For artificial intelligence to become a reality, it will need to make major breakthroughs in computational linguistics.

19.7 Cybernetics

The interdisciplinary field of cybernetics[18] began in the late 1940s when concepts such as information, feedback and regulation were generalized from engineering to other systems. These include systems such as living organisms, machines, robots and language. Norbert Wiener coined the term *cybernetics*, and it was taken from the Greek word 'κυβερνητη' (meaning steersman or governor). It is the study of communications and control and feedback in living organisms and machines to ensure efficient action.

The name is well chosen, as a steersman needs to respond to different conditions and feedback while steering a boat to travel to a particular destination. Similarly, the field of cybernetics is concerned with the interaction of goals, predictions, actions, feedback and responses in all kinds of systems. It uses models of organizations, feedback and goals to understand the capacity and limits of any system.

It is concerned with knowledge acquisition through control and feedback. Its principles are similar to human knowledge acquisition, where learning is achieved through a continuous process of feedback from parents and peers, which leads to adaptation and transformation of knowledge, rather than its explicit encoding.

The conventional belief in AI is that knowledge may be stored inside a machine and that the application of stored knowledge to the real world in this way constitutes intelligence. External objects are mapped to internal states on the machine, and machine intelligence is manifested by the manipulation of the internal states. This approach has been reasonably successful with rule-based expert systems, but it has made limited progress in creating intelligent machines. Therefore, alternative approaches such as cybernetics warrant further research. Cybernetics views information (or intelligence) as an attribute of an interaction, rather than something that is stored in a computer.

19.8 Logic and AI

Mathematical logic is used in the AI field to formalize knowledge and reasoning. Common-sense reasoning is required for solving problems in the real world, and McCarthy [Mc:59] argues that it is reasonable for logic to play a key role in the formalization of common-sense knowledge. This includes the formalization of basic facts about actions and their effects, facts about beliefs and desires and facts about knowledge and how it is obtained. His approach allows common-sense problems to be solved by logical reasoning.

Its formalization requires sufficient understanding of the common-sense world, and often the relevant facts to solve a particular problem are unknown. It may be that knowledge thought relevant may be irrelevant and vice versa. A computer may have millions of facts stored in its memory, and the problem is how to determine

[18] Cybernetics was defined by Couffignal (one of its pioneers) as the art of ensuring the efficacy of action.

which of these should be chosen from its memory to serve as premises in logical deduction.

McCarthy's influential 1959 paper discusses various common-sense problems such as getting home from the airport. Mathematical logic is the standard approach to express premises, and it includes rules of inferences that are used to deduce valid conclusions from a set of premises. Its rigorous deductive reasoning shows how new formulae may be logically deduced from a set of premises.

McCarthy's approach to programs with common sense has been criticized by Bar-Hillel and others on the grounds that common sense is fairly elusive and the difficulty that a machine would have in determining which facts are relevant to a particular deduction from its known set of facts.

Propositional calculus associates a truth value with each proposition and includes logical connectives to produce formulae such as $A \rightarrow B$, $A \wedge B$ and $A \vee B$. The truth values of the propositions are normally the binary values of *true* and *false*. There are other logics, such as three-valued logic or fuzzy logics that allow more than two truth values for a proposition. Predicate logic is more expressive than propositional logic and includes quantifiers and variables. It can formalize the syllogism 'All Greeks are mortal; Socrates is a Greek; therefore, Socrates is mortal'. The predicate calculus consists of:

- Axioms
- Rules for defining well-formed formulae
- Inference rules for deriving theorems from premises

A formula in predicate calculus is built up from the basic symbols of the language. These include variables, predicate symbols such as equality; function symbols; constants; logical symbols such as $\exists$, $\wedge$, $\vee$, $\neg$; and punctuation symbols such as brackets and commas. The formulae of predicate calculus are built from terms, where a *term* is defined recursively as a variable or individual constant or as some function containing terms as arguments. A formula may be an atomic formula or built from other formulae via the logical symbols.

There are several rules of inference associated with predicate calculus, and the most important of these are modus ponens and generalization. The rule of modus ponens states that given two formulae p and $p \rightarrow q$, then we may deduce q. The rule of generalization states that given a formula p, we may deduce $\forall(x)p$.

19.9 Computability, Incompleteness and Decidability

An algorithm (or procedure) is a finite set of unambiguous instructions to perform a specific task. The term 'algorithm' is named after the Persian mathematician *Al-Khwarizmi*. Church defined the concept of an algorithm formally in 1936, and he defined computability in terms of the lambda calculus. Turing defined computability in terms of the theoretical Turing machine. These formulations are equivalent.

Hilbert proposed *formalism* as a foundation for mathematics in the early twentieth century. A formal system consists of a *formal language*, a set of axioms and rules of *inference*. Hilbert's program was concerned with the formalization of mathematics (i.e. the axiomatization of mathematics) together with a proof that the axiomatization was consistent. Its goals were to:

- Develop a formal system where the truth or falsity of any mathematical statement may be determined.
- A proof that the system is consistent (i.e. that no contradictions may be derived).

A proof in a formal system consists of a sequence of formulae, where each formula is either an axiom or derived from one or more preceding formulae in the sequence by one of the rules of inference. Hilbert believed that every mathematical problem could be solved, and he therefore expected that the formal system of mathematics would be *complete* (i.e. all truths could be proved within the system) and *decidable*: i.e. that the truth or falsity of any mathematical proposition could be determined by an algorithm. However, Church and Turing independently showed this to be impossible in 1936, and the only way to determine whether a statement is true or false is to try to solve it.

Russell and Whitehead published *Principia Mathematica* in 1910, and this three-volume work on the foundations of mathematics attempted to derive all mathematical truths in arithmetic from a well-defined set of axioms and rules of inference. The questions remained whether the Principia was *complete* and *consistent*. That is, is it possible to derive all the truths of arithmetic in the system and is it possible to derive a contradiction from the Principia's axioms?

Gödel's second incompleteness theorem [Goe:31] showed that first-order arithmetic is incomplete and that the consistency of first-order arithmetic cannot be proved within the system. Therefore, if first-order arithmetic cannot prove its own consistency, then it cannot prove the consistency of any system that contains first-order arithmetic. These results dealt a fatal blow to Hilbert's program.

19.10 Robots

The first use of the term 'robot' was by the Czech playwright Karel Capek in his play *Rossum's Universal Robots* performed in Prague in 1921. The word 'robot' is from the Czech word for forced labour. The theme explored is whether it is ethical to exploit artificial workers in a factory and how the robots should respond to their exploitation. Capek's robots were not mechanical or metal in nature and were instead created through chemical means.

Asimov wrote several stories about robots in the 1940s including the story of a robotherapist. He predicted the rise of a major robot industry, and he also introduced a set of rules (or laws) for good robot behaviour. These are known as the Three Laws of Robotics (Table 19.1), and Asimov later added a fourth law.

Table 19.1 Laws of Robotics

Law	Description
Law zero	A robot may not injure humanity or, through inaction, allow humanity to come to harm
Law one	A robot may not injure a human being or, through inaction, allow a human being to come to harm, unless this would violate a higher order law
Law two	A robot must obey orders given it by human beings, except where such orders would conflict with a higher-order law
Law three	A robot must protect its own existence as long as such protection does not conflict with a higher-order law

The term 'robot' is defined by the Robot Institute of America as:

Definition 19.1 (Robots) *A re-programmable, multifunctional manipulator designed to move material, parts, tools, or specialized devices through various programmed motions for the performance of a variety of tasks.*

Joseph Engelberger and George Devol are considered the fathers of robotics, and they set up the first manufacturing company 'Unimation' to make robots. Their first robot was called the 'Unimate'. These robots were very successful and reliable and saved their customer (General Motors) money by replacing staff with machines.

Robots are very effective at doing clearly defined repetitive tasks, and there are many sophisticated robots in the workplace today. The robot industry plays a major role in the automobile sector, and these are mainly industrial manipulators that are essentially computer controlled 'arms and hands'. However, fully functioning androids are many years away.

Robots may also improve the quality of life for workers, as they can free human workers from performing dangerous or repetitive tasks. They consistently produce ($24 \times 7 \times 365$) high-quality products at a low cost to consumers. They will, of course, from time to time require servicing by engineers or technicians. However, there are impacts on workers whose jobs are displaced by robots.

19.11 Neural Networks

The term 'neural network' refers to an interconnected group of processing elements called nodes or neurons. These neurons cooperate and work together to produce an output function. Neural networks may be artificial or biological. A biological network is part of the human brain, whereas an artificial neural network is designed to mimic some properties of a biological neural network. The processing of information by a neural network is done in parallel rather than in series.

A unique property of a neural network is fault tolerance: i.e. it can still perform (within certain tolerance levels) its overall function even if some of its neurons are not functioning. Neural network may be trained to learn to solve complex problems from a set of examples. These systems may also use the acquired knowledge to generalize and solve unforeseen problems.

A biological neural network is composed of billions of neurons (or nerve cells). A single neuron may be physically connected to thousands of other neurons, and the total number of neurons and connections in a network may be enormous. The human brain consists of many billions of neurons, and these are organized into a complex intercommunicating network. The connections are formed through *axons*[19] to *dendrites*,[20] and the neurons can pass electrical signals to each other. These connections are not just the binary digital signals of *on* or *off*, and instead the connections have varying strength, which allows the influence of a given neuron on one of its neighbours to vary from very weak to very strong.

That is, each connection has an individual weight (or number) associated with it that indicates its strength. Each neuron sends its output value to all other neurons to which it has an outgoing connection. The output of one neuron can influence the activations of other neurons causing them to fire. The neuron receiving the connections calculates its activation by taking a *weighted sum* of the input signals. Networks learn by changing the weights of the connections. Many aspects of brain function, especially the learning process, are closely associated with the adjustment of these connection strengths. Brain activity is represented by particular patterns of firing activity among the network of neurons. This simultaneous cooperative behaviour of a huge number of simple processing units is at the heart of the computational power of the human brain.[21]

Artificial neural networks aim to simulate various properties of biological neural networks. They consist of many hundreds of simple processing units, which are wired together in a complex communication network. Each unit or node is a simplified model of a real neuron which fires[22] if it receives a sufficiently strong input signal from the other nodes to which it is connected. The strength of these connections may be varied in order for the network to perform different tasks corresponding to different patterns of node firing activity. The objective is to solve a particular problem, and artificial neural networks have been applied to *speech recognition problems*, *image analysis* and so on.

The human brain employs massive parallel processing, whereas artificial neural networks have provided simplified models of the neural processing that takes place in the brain. The largest artificial neural networks are tiny compared to biological neural networks. The challenge for the field is to determine what properties individual neural elements should have to produce something useful representing intelligence.

---

[19] These are essentially the transmission lines of the nervous system. They are microscopic in diameter and conduct electrical impulses. The axon is the output from the neuron and the dendrites are input.

[20] Dendrites extend like the branches of a tree. The origin of the word dendrite is from the Greek word (δενδρον) for tree.

[21] The brain works through massive parallel processing.

[22] The firing of a neuron means that it sends off a new signal with a particular strength (or weight).

Neural networks differ from the traditional von Neumann architecture, which is based on the sequential execution of machine instructions. The origins of neural networks lie in the attempts to model information processing in biological systems. This relies more on parallel processing as well as on implicit instructions based on pattern recognition from sense perceptions of the external world.

The nodes in an artificial neural network are composed of many simple processing units, which are connected into a network. Their computational power depends on working together (parallel processing) on any task, and computation is related to the dynamic process of node firings rather than sequential execution of instructions. This structure is much closer to the operation of the human brain and leads to a computer that may be applied to a number of complex tasks.

19.12 Expert Systems

An expert system is a computer system that contains domain knowledge of one or more human experts in a narrow specialized domain. It consists of a set of rules (or knowledge) supplied by the domain experts about a specific class of problems and allows knowledge to be stored and intelligently retrieved. The effectiveness of the expert system is largely dependent on the accuracy of the rules provided, as incorrect inferences will be drawn with incorrect rules. Several commercial expert systems have been developed since the 1960s.

Expert systems have been a success story in the AI field. They have been applied to the medical field, equipment repair and investment analysis. They employ a logical reasoning capability to draw conclusions from known facts, as well as recommending an appropriate course of action to the user. An expert system consists of the following components (Table 19.2): a knowledge base, an inference engine, an explanatory facility, a user interface and a database.

Human knowledge of a specialty is of two types: namely, public knowledge and private knowledge. The former includes the facts and theories documented in textbooks and publications, whereas the latter refers to knowledge that the expert possesses that has not found its way into the public domain. The latter often consists of rules of thumb or heuristics that allow the expert to make an educated guess where required, as well as allowing the expert to deal effectively with incomplete or erroneous data. It is essential that the expert system encodes both public and private knowledge to enable it to draw valid inferences.

Table 19.2 Expert systems

Component	Description
Knowledge base	The knowledge base is represented as a set of rules of the form (*if* condition, *then* action)
Inference engine	Carries out reasoning by which expert system reaches conclusion
Explanatory facility	Explains how a particular conclusion was reached
User interface	Interface between user and expert system
Database/memory	Set of facts used to match against IF conditions in knowledge base

The inference engine is made up of many inference rules that are used by the engine to draw conclusions. Rules may be added or deleted without affecting other rules, and this reflects the normal updating of human knowledge. Out-of-date facts may be deleted, as they are no longer used in reasoning, while new knowledge may be added and applied in reasoning. The inference rules use reasoning that is closer to human reasoning, and the two main types of reasoning are backward chaining and forward chaining. Forward chaining starts with the data available and uses the inference rules to draw intermediate conclusions until a desired goal is reached. Backward chaining starts with a set of goals and works backwards to determine if one of the goals can be met with the data that is available.

The expert system makes its expertise available to decision-makers who need answers quickly. This is extremely useful as often there is a shortage of experts, and the availability of an expert computer with in-depth knowledge of specific subjects is therefore very attractive. Expert systems may also assist managers with long-term planning. There are many small expert systems available that are quite effective in a narrow domain.

Several expert systems (e.g. Dendral, Mycin, and Colossus) have been developed. Dendral (Dendritic Algorithm) was developed at Stanford University in the mid-1960s, and its objectives were to assist the organic chemist with the problem of identifying unknown organic compounds and molecules by computerized spectrometry. This involved the analysis of information from mass spectrometry graphs and knowledge of chemistry. Dendral automated the decision-making and problem-solving process used by organic chemists to identify complex unknown organic molecules. It was written in LISP and it showed how an expert system could employ rules, heuristics and judgement to guide scientists in their work.

Mycin was developed at Stanford University in the 1970s. It was written in LISP and was designed to diagnose infectious blood diseases and to recommend appropriate antibiotics and dosage amounts corresponding to the patient's body weight. It had a knowledge base of approximately 500 rules and a fairly simple inference engine. Its approach was to query the physician running the program with a long list of yes/no questions. Its output consisted of various possible bacteria that could correspond to the blood disease, along with an associated probability that indicated the confidence in the diagnosis. It also included the rationale for the diagnosis and a course of treatment appropriate to the diagnosis.

Mycin had a correct diagnosis rate of 65 %. This was better than the diagnosis of most physicians who did not specialize in bacterial infections. However, its diagnosis rate was less than that of experts in bacterial infections who had a success rate of 80 %. Mycin was never actually used in practice due to legal and ethical reasons on the use of expert systems in medicine. For example, if the machine makes the wrong diagnosis, who is to be held responsible?

Colossus was an expert system used by several Australian insurance companies. It was used to help insurance adjusters assess personal injury claims and helped to improve consistency, objectivity and fairness in the claims process. It guides the adjuster through an evaluation of medical treatment options, the degree of pain and suffering of the claimant and the extent that there is permanent impairment to the claimant, as well as the impact of the injury on the claimant's lifestyle. The Computer Sciences Corporation (CSC) developed it.

19.13 Review Questions

1. Discuss Descartes and his rationalist philosophy and his relevance to artificial intelligence.
2. Discuss the Turing Test and its relevance on strong AI.
3. Discuss Searle's Chinese room rebuttal arguments. What are your views on Searle's argument?
4. Discuss the philosophical problems underlying artificial intelligence.
5. Discuss the applicability of logic to artificial intelligence.
6. Discuss neural networks and their applicability to artificial intelligence.
7. Discuss expert systems and their applications to the AI field.
8. Discuss the applications of cybernetics to the AI field.
9. Discuss the applications of phenomenology to the AI field.

19.14 Summary

Artificial intelligence is a multidisciplinary field, and its branches include logic, philosophy, psychology, linguistics, machine vision, neural networks and expert systems. Turing believed that machine intelligence was achievable, and he devised the 'Turing Test' to judge if a machine was intelligent. Searle's Chinese room argument is a rebuttal of strong AI, and it aims to demonstrate that a machine will never have the same cognitive qualities as a human even if it passes the Turing Test.

McCarthy proposed programs with common-sense knowledge and reasoning formalized with logic. He argued that human-level intelligence could be achieved with a logic-based system. Cognitive psychology is concerned with cognition and some of its research areas include perception, memory, learning, thinking, logic and problem solving. Linguistics is the *scientific* study of *language* and includes the study of syntax and semantics.

Artificial neural networks aim to simulate various properties of biological neural networks. They consist of many hundreds of simple processing units that are wired together in a complex communication network. Each unit or node is a simplified model of a real neuron which fires if it receives a sufficiently strong input signal from the other nodes to which it is connected. The strength of these connections may be varied in order for the network to perform different tasks corresponding to different patterns of node firing activity.

An expert system is a computer system that allows knowledge to be stored and intelligently retrieved. It is a program that is made up of a set of rules (or knowledge). The domain experts generally supply the rules about a specific class of problems. Expert systems include a problem-solving component that allows an analysis of the problem, as well as recommending an appropriate course of action to solve the problem.

History of Databases

20

Abstract
We present a short history of databases including the hierarchical model, the network model and the relational model. We discuss the relational model as developed by Codd at IBM in more detail, as most databases used today are relational. There is a short discussion on the SQL and on the Oracle database.

Key Topics
Hierarchical model
Network model
Relational model
Table
Key
Index
SQL
Oracle database

20.1 Introduction

A database (DB) is essentially an organized collection of data and consists of schemas, tables, queries, reports and views. It is organized in such a way that a computer program (termed the database management system) may easily select and analyse the desired pieces of data. A database holds information about many different types of entities, as well as information about the relationships between the entities.

A database management system (DBMS) is a collection of software programs that allows a user to store, modify and extract data from a database. The interaction

G. O'Regan, *Introduction to the History of Computing*, Undergraduate Topics in Computer Science, DOI 10.1007/978-3-319-33138-6_20

between the users and the database is through the DBMS, and it enables the definition, creation, query, update and administration of databases. There are three main categories of database management systems, and these are hierarchical, network and relational models. These differ in how the DBMS organizes data internally, and this determines the speed and efficiency of data retrieval from the database.

A network model database is perceived by the user to be a collection of record types and relationships between these record types organized as a network. The network model defines the relationships explicitly as part of the structure of the network. A hierarchical model is perceived by a user to be a collection of hierarchies or trees, and it is a more restricted structure than the network model as only one arrow may enter each box on the network. A relational model is perceived by the user to be a collection of tables (or relations), and it has been the most popular category of databases since the 1980s.

Early work on database management systems began in the 1960s as part of the Apollo mission to land man on the moon. It was clear that the existing systems were not capable of handling the coordination of the vast amounts of data required for the project. IBM developed the Generalized Update Access Method (GUAM) product in 1964, and this product evolved into Data Language/1 (DL/1). DL/1 is the data management component of the information management system (IMS) database, which was one of the earliest database management systems when it was introduced in 1968. IMS used the hierarchical model.

The CODASYL committee[1] set up a database task group and devised a standard, which became known as the 'CODASYL approach'. This became the network standard, and it was defined in the late 1960s. The standard was introduced in 1971.

Codd proposed the relational model, a radically new approach to the management of data in 1970, and IBM developed the prototype system called System R in the 1970s. Commercial relational database systems were introduced from the early 1980s, and today relational databases are much more widely used than network or hierarchical databases. Among the popular relational databases used today are Oracle, Microsoft SQL Server and Informix.

20.2 Hierarchical and Network Models

A database management system uses the network model if the data relationships are defined in terms of a graph. The relationships are defined in terms of records (a record is a collection of fields, with each field containing one value), which are connected together via links. Any given record may have several parent records and several dependent records. Cycles are permitted in the model. Charles Bachman and others on the CODASYL committee defined the network model in the late 1960s.

[1] The CODASYL committee is the group that defined and standardized the COBOL programming language. It was also involved in work in standardizing database interfaces.

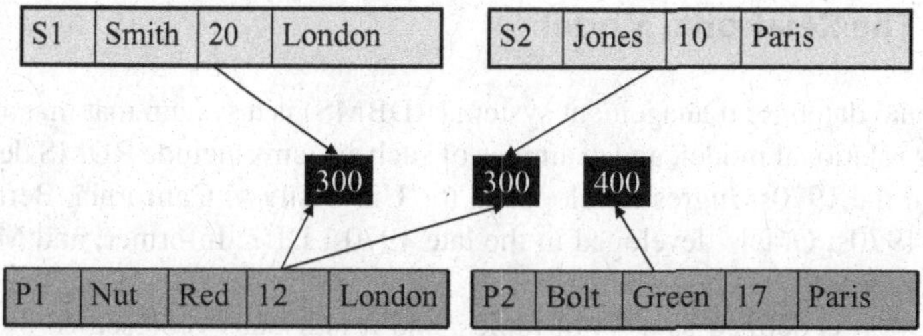

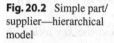

Fig. 20.1 Simple part/supplier—network model

Fig. 20.2 Simple part/supplier—hierarchical model

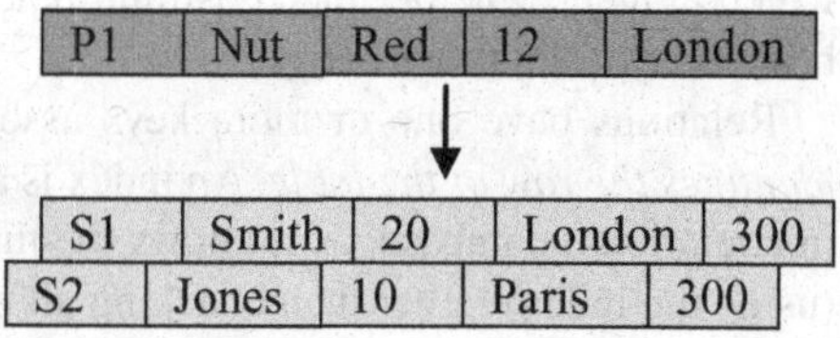

Among the well-known databases that were based on the network model was General Electric's Integrated Data Store (IDS) and the Integrated Database Management System (IDMS). Both of these mainframe databases were introduced in the early 1970s.

For a possible network view of suppliers and parts, the data would be presented in a simple graph-like structure (Fig. 20.1), which allows many-to-many relationships to be expressed. For more detailed information, see [Dat:81].

A database management system uses the hierarchical model if the data relationships are defined in terms of hierarchies (i.e. in a tree-like structure). The relationships are simple but inflexible (as they are one to many). The data are defined as records, which are connected to each other through links. Each child record may have only one parent, whereas each parent record may have several children records. The whole tree (starting from the root) needs to be traversed in order to retrieve data from a hierarchical database. In other words, the hierarchical model is a more restricted version of the network model, where no box can have more than one arrow entering the box although several arrows can leave a box.

For a possible hierarchical view of suppliers and parts, the data would be presented in a simple tree-like structure (Fig. 20.2). Each tree consists of one part record together with a set of supplier record occurrences, one for each supplier of the part. For more detailed information, see [Dat:81].

The database access and manipulation component of the hierarchical model is termed Data Language/1, and it includes a data definition language and a data manipulation language. The IBM Information Management System (IMS) is one of the most widely used hierarchical databases, and it was created in the late 1960s.

20.3 The Relational Model

A relational database management system (RDBMS) is a system that manages data using the relational model, and examples of such systems include RDMS developed at MIT in the 1970s; Ingres developed at the University of California, Berkeley, in the mid-1970s; Oracle developed in the late 1970s; DB2; Informix; and Microsoft SQL Server.

A relation is defined as a set of tuples, and it is usually represented by a table. A table is data organized in rows and columns, with the data in each column of the table of the same data type. Constraints may be employed to provide restrictions on the kinds of data that may be stored in the relations. These are Boolean expressions which indicate whether the constraint holds or not and are a way of implementing business rules in the database.

Relations have one or more keys associated with them, and the *key uniquely identifies the row of the table*. An index is a way of providing fast access to the data in a relational database, as it allows the tuple in a relation to be looked up directly (using the index) rather than checking all of the tuples in the relation.

The structured query language (SQL) is a computer language that tells the relational database what to retrieve and how to display it. A stored procedure is executable code that is associated with the database, and it is used to perform common operations on the database.

The concept of a relational database was first described in a paper *A Relational Model of Data for Large Shared Data Banks* by Codd [Cod:70]. A relational database is a database that conforms to the *relational model*, and it may be defined as a set of *relations* (or tables).

Codd (Fig. 20.3) was a British mathematician, computer scientist and IBM researcher, who initially worked on the SSEC (Selective Sequence Electronic Calculator) project in New York and then on the IBM 701 and 702 computers. He later worked on the IBM 7030 Stretch computer (IBM's first transistorized computer). He was the creator of STEM (statistical database expert manager).

He developed the *relational database model* in the late 1960s, and he published an internal IBM paper on the relational model in 1969. Today, this is the standard way that information is organized and retrieved from computers, and relational databases are at the heart of systems from hospitals' patient records to airline flight and schedule information.

IBM was promoting its IMS hierarchical database in the 1970s, and it showed little interest or enthusiasm for Codd's new relational database model. It made business sense for IBM to preserve revenue for the IMS/DB model, rather than embarking on a new technology. However, IBM agreed to implement Codd's ideas on the relational model for the *System R research project* in the 1970s, and this project demonstrated the power of the model, as well as demonstrating good transaction processing performance. The project introduced a data query language that was initially called SEQUEL (later renamed to SQL), and this language was designed to retrieve and manipulate data in the IBM database.

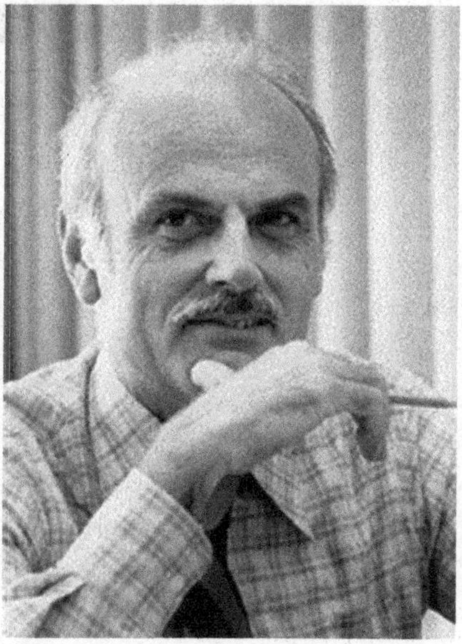

Fig. 20.3 Edgar Codd

Codd continued to develop and extend his relational model, and several theorems are named after him. In later years he proposed a three-valued logic to deal with missing or undefined information and even proposed a four-valued logic in the 1990s. These proposals were never implemented and were controversial at the time. The relational model became popular from the early 1980s, and Codd received the ACM Turing Award in 1981 for his development of the relational database model.

A binary relation R(A,B) where A and B are sets is a subset of the Cartesian product (A × B) of A and B. The domain of the relation is A, and the codomain of the relation is B. The notation *a*R*b* signifies that there is a relation between *a* and *b* and that $(a,b) \in R$. An *n*-ary relation R $(A_1, A_2, \ldots A_n)$ is a subset of the Cartesian product of the *n* sets: i.e. a subset of $(A_1 \times A_2 \times \ldots \times A_n)$. However, an *n*-ary relation may also be regarded as a binary relation R(A,B) with $A = A_1 \times A_2 \times \ldots \times A_{n-1}$ and $B = A_n$.

The data in the relational model are represented as a mathematical *n*-ary relation. In other words, a relation is defined as a set of *n-tuples* and is usually represented by a table. A table is a visual representation of the relation, and the data is organized in *rows* and *columns*. The data stored in each column of the table is of the same *data type*.

The basic relational building block is the domain or data type (often called just type). Each row of the table represents one *n*-tuple (one tuple) of the relation, and the number of tuples in the relation is the cardinality of the relation. Consider the PART relation taken from [Dat:81], where this relation consists of a heading and the body. There are five data types representing part numbers, part names, part colours, part weights and locations in which the parts are stored. The body consists of a set of *n*-tuples. The PART relation is of cardinality 6 (Fig. 20.4).

P#	PName	Colour	Weight	City
P1	Nut	Red	12	London
P2	Bolt	Green	17	Paris
P3	Screw	Blue	17	Rome
P4	Screw	Red	14	London
P5	Cam	Blue	12	Paris
P6	Cog	Red	19	London

Fig. 20.4 PART relation

```
DOMAIN  PART_NUMBER      CHARACTER(6)
DOMAIN  PART_NAME        CHARACTER(20)
DOMAIN  COLOUR           CHARACTER(6)
DOMAIN  WEIGHT           NUMERIC(4)
DOMAIN  LOCATION         CHARACTER(15)

RELATION PART
        (P#              : DOMAIN  PART_NUMBER
        PNAME            : DOMAIN  PART_NAME
        COLOUR           : DOMAIN  COLOUR
        WEIGHT           : DOMAIN  WEIGHT
        CITY             : DOMAIN  LOCATION)
```

Fig. 20.5 Domains vs. attributes

Strictly speaking there is no ordering defined among the tuples of a relation, since a relation is a set and sets are not ordered. However, in practice, relations are often considered to have an ordering.

There is a distinction between a domain and the columns (or attributes) that are drawn from that domain. An *attribute* represents the *use* of a domain within a relation, and the distinction is often emphasized by giving attributes names that are distinct from the underlying domain. The difference between domains and attributes can be seen in the PART relation (Fig. 20.5) from [Dat:81].

A *normalized relation* satisfies the property that at every row and column position in the table, there is exactly one value (i.e. never a set of values). All relations in a relational database are required to satisfy this condition, and an un-normalized relation may be converted into an equivalent normalized form.

It is often the case that within a given relation, there is one attribute with values that is unique within the relation and can thus be used to identify the tuples of the relation. For example, the attribute P# of the PART relation has this property since each PART tuple contains a distinct P# value, which may be used to distinguish that tuple from all other tuples in the relation. P# is termed the *primary key* for the PART relation. A candidate key that is not the primary key is termed the *alternate key*.

An index is a way of providing quicker access to the data in a relational database, as it allows the tuple in a relation to be looked up directly (using the index) rather than checking all of the tuples in the relation.

The consistency of a relational database is enforced by a set of constraints that provide restrictions on the kinds of data that may be stored in the relations. The constraints are declared as part of the logical schema and are enforced by the database management system. They are used to implement the business rules for the database.

20.4 Structured Query Language (SQL)

Codd proposed the Alpha language as the database language for his relational model. However, IBM's implementation of his relational model in the System R project introduced a data query language that was initially called SEQUEL (later renamed to SQL). This language did not adhere to Codd's relational model but became the most popular and widely used database language. It was designed to retrieve and manipulate data in the IBM database, and its operations included *insert, delete, update, query*, schema creation and modification and data access control.

Structured query language (SQL) is a computer language that tells the relational database what to retrieve and how to display it. It was designed and developed at IBM by Donald Chamberlin and Raymond Boyce, and it became an ISO standard in 1987.

The most common operation in SQL is the query command, which is performed with the SELECT statement. The SELECT statement retrieves data from one or more tables, and the query specifies one or more columns to be included in the result. Consider the example of a query that returns a list of expensive books (defined as books that cost more than 100.00):

```
SELECT *[2]
   FROM  Book
   WHERE Price > 100.00
   ORDER by title;
```

The *data manipulation language* (DML) is the subset of SQL used to add, update and delete data. It includes the INSERT, UPDATE and DELETE commands. The *data definition language* (DDL) manages table and index structure and includes the CREATE, ALTER, RENAME and DROP statements.

There are extensions to standard SQL that add programming language functionality. A stored procedure is executable code that is associated with the database. It is usually written in an imperative programming language, and it is used to perform common operations on the database.

[2] The asterisk (*) indicates that all columns of the Book table should be included in the result.

Oracle is recognized as a world leader in relational database technology, and its products play a key role in business computing. An Oracle database consists of a collection of data managed by an Oracle database management system. Today, Oracle is the main standard for database technology, and it is described in the next section.

20.5 Oracle Database

An Oracle database is a collection of data treated as a unit, and the database is used to store and retrieve related information. The database server manages a large amount of data in a multi-user environment. It allows concurrent access to the data, and the database management system prevents unauthorized access to the database. It also provides a smooth recovery of database information in the case of an outage or any other disruptive event.

Every Oracle database consists of one or more physical data files, which contain all of the database data, and a control file that contains entries that specify the physical structure of the database.

An Oracle database includes logical storage structures that enable the database to have control of disc space use. A schema is a collection of database objects, and the schema objects are the logical structures that directly refer to the database's data. They include structures such as tables, views and indexes.

Tables are the basic unit of data storage in an Oracle database, and each table has several rows and columns. An index is an optional structure associated with a table, and it is used to enhance the performance of data retrieval. The index provides an access path to the table data. A view is the customized presentation of data from one or more tables. It does not contain actual data and derives the data from the actual tables on which it is based.

Each Oracle database has a data dictionary, which stores information about the logical and physical structure of the database. The data dictionary is created when the database is created and is updated automatically by the Oracle database to ensure that it accurately reflects the status of the database at all times.

An Oracle database uses memory structures and various processes to manage and access the database. These include server processes, background processes and user processes.

A database administrator (DBA) is responsible for setting up the Oracle database server and application tools. This role is concerned with allocating system storage and planning future storage requirements for the database management system. The DBA will create appropriate storage structures to meet the needs of application developers who are designing a new application. The access to the database will be monitored and controlled, and the performance of the database monitored and optimized. The DBA will plan backups and recovery of database information.

20.6 Review Questions

1. What is a database?
2. What is a database management system?
3. Explain the differences between relational, hierarchical and network databases.
4. Explain the difference between a key and an index.
5. What is a stored procedure?
6. What is the role of the Oracle DBA?
7. Explain the differences between tables, views and schemas.
8. What is SQL?.
9. What is an Oracle database?

20.7 Summary

A database (DB) is essentially a collection of data organized in such a way that a computer program may easily select the desired pieces of data. A database management system (DBMS) is a collection of software programs that allows a user to store, modify and extract data from a database.

There are three main categories of database management systems, and these are hierarchical, network and relational models. A network model database is perceived by the user to be a collection of record types and relationships between them organized as a network. A hierarchical model is perceived by a user to be a collection of hierarchies or trees, and it is a more restricted structure than the network model. A relational model is perceived by the user to be a collection of tables (or relations).

Early work on database management systems began in the 1960s, and IBM developed the information management system (IMS) database in the late 1960s. This hierarchical database was one of the earliest database management systems.

Codd proposed the relational model as a new approach to the management of data in 1970, and IBM developed the prototype System R relational database in the 1970s. Relational databases are now dominant with the hierarchical and network model mainly of historical interest.

Glossary

ABC	Atanasoff-Berry Computer
ACS	Advanced Computing Systems
AI	Artificial Intelligence
ALGOL	Algorithmic language
AMD	Advanced Micro Devices
AMPS	Advanced Mobile Phone System
ANS	Advanced Network Services
ANSI	American National Standards Institute
API	Application Programming Interface
ARPA	Advanced Research Projects Agency
ASCC	Automatic Sequence Controlled Calculator
ASCII	American Standard Code for Information Interchange
AXE	Automatic Exchange Electric switching system
B2B	Business to Business
B2C	Business to Consumer
BASIC	Beginners All-purpose Symbolic Instruction Code
BBN	Bolt, Beranek and Newman
BCS	British Computer Society
BIOS	Basic Input Output System
CD	Compact Disc
CDC	Control Data Corporation
CDMA	Code Division Multiple Access
CEO	Chief Executive Officer
CERN	Conseil Européen pour la Recherche Nucléaire
CERT	Certified Emergency Response Team
CMM®	Capability Maturity Model
CMMI®	Capability Maturity Model Integration
CMS	Conversational Management System
COBOL	Common Business Oriented Language
CODASYL	Conference on Data Systems Languages
COPQ	Cost of Poor Quality
COTS	Customized Off The Shelf
CP/M	Control Program for Microcomputers

G. O'Regan, *Introduction to the History of Computing*, Undergraduate Topics in Computer Science, DOI 10.1007/978-3-319-33138-6

CPU	Central Processing Unit
CSIRAC	Council for Scientific and Industrial Research Automatic Computer
CRT	Cathode Ray Tube
CTSS	Compatible Time-Sharing System
DARPA	Defense Advanced Research Project Agency
DB	Database
DBA	Database Administrator
DBMS	Database Management System
DDL	Data Definition Language
DEC	Digital Equipment Corporation
DL/1	Data Language 1
DML	Data Manipulation Language
DNS	Domain Naming System
DOS	Disk Operating System
DRAM	Dynamic Random Access Memory
DRI	Digital Research Incorporated
DSDM	Dynamic Systems Development Method
DVD	Digital Versatile Disc
EDSAC	Electronic Delay Storage Automatic Calculator
EDVAC	Electronic Discrete Variable Automatic Computer
EMCC	Eckert-Mauchly Computer Corporation
ENIAC	Electronic Numerical Integrator and Computer
ETH	Eidgenössische Technische Hochschule
ETACS	Extended TACs
ETSI	European Telecommunications Standards Institute
FAA	Federal Aviation Authority
FDMA	Frequency Division Multiple Access
FTP	File Transfer Protocol
GB	Gigabyte
GECOS	General Electric Comprehensive Operating System
GL	Generation Language
GPRS	General Packet Radio Service
GSM	Global System Mobile
GUAM	Generalised Update Access Method
GUI	Graphical User Interface
HP	Hewlett Packard
HTML	Hypertext Markup Language
HTTP	Hypertext Transport Protocol
IBM	International Business Machines
IC	Integrated Circuit
ICBM	Intercontinental Ballistic Missile
IDE	Integrated Development Environment
IDMS	Integrated Database Management System
IDS	Integrated Data Store
IE	Internet Explorer

IEEE	Institute of Electrical and Electronic Engineers
IMP	Interface Message Processor
IMS	Information Management System
INWG	International Network-Working Group
IOS	Internetwork operating system
IP	Internet Protocol
IPCS	Interactive Problem Control System
IPO	Initial Public Offering
ISEB	Information Systems Examination Board
ISO	International Standards Organization
IT	Information Technology
JAD	Joint Application Development
JCL	Job Control Language
JVM	Java Virtual Machine
KB	Kilobyte
KLOC	Thousand Lines of Code
LAN	Local Area Network
LED	Light Emitting Diode
LEO	Lyons Electronic Office
LEO	Low Earth Orbit
LSI	Large Scale Integration
MADC	Manchester Automatic Digital Computer
MB	Megabyte
ME	Millennium
MFT	Multiple Programming with a Fixed number of Tasks
MIDI	Musical Instrument Digital Interface
MIT	Massachusetts Institute of Technology
MITS	Micro Instrumentation and Telemetry System
MOS	Metal Oxide Semiconductor
MSI	Medium Scale Integration
MS/DOS	Microsoft Disk Operating System
MTX	Mobile Telephone Exchange
MVS	Multiple Virtual Storage
MVT	Multiple Programming with a Variable number of Tasks
NAP	Network Access Point
NASA	National Aeronautics and Space Administration
NATO	North Atlantic Treaty Organization
NCP	Network Control Protocol
NMT	Nordic Mobile Telephony system
NORAD	North American Aerospace Defense
NPL	National Physical Laboratory
NR	Norwegian Research
NSF	National Science Foundation
OS	Operating System
PARC	Palo Alto Research Centre

PC	Personal Computer
PC/DOS	Personal Computer Disk Operating System
PDA	Personal Digital Assistant
PDP	Programmed Data Processor
PL/M	Programming Language for Microcomputers
PTT	Postal Telephone and Telegraph
RAD	Rapid Application Development
RAM	Random Access Memory
RDBMS	Relational Database Management System
RIM	Research in Motion
ROM	Read Only Memory
RSCS	Remote Spooling Communications Subsystem
RUP	Rational Unified Process
SAGE	Semi-Automatic Ground Environment
SECD	Stack, Environment, Control, Dump
SEI	Software Engineering Institute
SID	Sound Interface Device
SIM	Subscriber Identity Module
SMS	Short Message Service
SMTP	Simple Mail Transfer Program
SNS	Social Networking Site
SPREAD	System Programming, Research, Engineering and Design
SQL	Structured Query Language
SRI	Stanford Research Institute
SSEM	Small Scale Experimental Machine
SSI	Small Scale Integration
SSL	Secure Socket Layer
SWF	Small Web Format
TACS	Total Access Communication
TCP	Transport Control Protocol
TI	Texas Instrument
TSO	Time Sharing Option
UAT	User Acceptance Testing
UCLA	University of California (Los Angeles)
UDP	User Datagram Protocol
ULSI	Ultra Large Scale Integration
UML	Unified Modelling Language
UNIVAC	Universal Automatic Computer
URL	Universal Resource Locator
VAX	Virtual Address eXtension
VDM	Vienna Development Method
VLSI	Very Large Scale Integration
VM	Virtual Memory
VMS	Virtual Memory System
W3C	World Wide Web Consortium
WCDMA	Wideband CDMA

References

[AnDa:14] Thomas A, Michael D (2014) Operating systems: principles and practice. Recursive Books, West Lake Hills

[AnL:95] Anglin WS, Lambek J (1995) The heritage of Thales. Springer, New York

[Bab:42] Menabrea LF (1842) Sketch of the analytic engine. Invented by Charles Babbage (trans: Lada Ada Lovelace LA). Bibliothèque Universelle de Genève

[Bag:12] Bagnall B (2012) Commodore: a company on the edge, 2nd edn. Variant Press, Winnipeg

[Bec:00] Beck K (2000) Extreme programming explained embrace change. Addison Wesley, Reading

[Ber:99] George B (1999) Principles of human knowledge. Oxford University Press, Oxford. (Originally published in 1710)

[BL:00] Berners-Lee T (2000) Weaving the web. Collins Book, New York

[Blo:04] The man who could have been Bill Gates. Bloomberg Business Week Magazine. October 2004

[Boe:88] Barry B (1988) A spiral model for software development and enhancement. Computer 21:61–72

[Boo:48] George B (1848) The calculus of logic. Camb Dublin Math J III:183–198

[Boo:58] George B (1958) An investigation into the laws of thought. Dover Publications, New York. (First published in 1854)

[Boy:04] Boyer C (2004) The 360 revolution. IBM

[Brk:75] Fred B (1975) The mythical man month. Addison Wesley, Reading

[Brk:86] Fred B (1986) No silver bullet. Essence and accidents of software engineering. In: Information processing. Elsevier, Amsterdam

[Bus:45] Bush V (1945) As we may think. The Atlantic Monthly 176(1):101–108

[Bux:75] Buxton JN, Naur P, Randell B (1975) Software engineering. Petrocelli. Report on two NATO Conferences held in Garmisch, Germany (October 1968) and Rome, Italy (October 1969)

[By:94] Halfhill T (1994) R.I.P. Commodore. 1954–1994. A look at an innovative industry pioneer, whose achievements have been largely forgotten. Byte Magazine, August 1994

[ChR:02] Henry C, Richard R (2002) The role of the business model in capturing value from innovation: evidence from xerox corporation's technology spin-off companies. Ind Corp Chang 11(3):529–555

[CKS:11] Chrissis MB, Mike C, Sandy S (2011) CMMI. Guidelines for process integration and product improvement, 3rd edn, SEI series in software engineering. Addison Wesley, Upper Saddle River

[Cod:70] Codd EF (1970) A relational model of data for large shared data banks. Commun ACM 13(6):377–387

[Dat:81] Date CJ (1981) An introduction to database systems, 3rd edn, The systems programming series. Addison-Wesley, Reading

G. O'Regan, *Introduction to the History of Computing*, Undergraduate Topics in Computer Science, DOI 10.1007/978-3-319-33138-6

[Dei:90] Deitel HM (1990) Operating systems, 2nd edn. Addison Wesley.
[Des:99] Descartes R (1999) Discourse on method and meditations on first philosophy, 4th edn. Translated by Cress D. Hackett Publishing Company, Indianapolis
[Dij:68] Dijkstra EW (1968) Go to statement considered harmful. Commun ACM 51:7–9
[Dij:72] Dijkstra EW (1972) Structured programming. Academic Press, London\New York
[Edw:11] Edwards B (2011) The history of atari computers. PC World.
[Fag:76] Fagan M (1976) Design and code inspections to reduce errors in software development. IBM Syst J 15(3):182–210
[Fer:03] Georgina F (2003) A computer called LEO: lyons tea shops and the world's first office computer. Fourth Estate Ltd, London
[Ger:13] Jon G (2013) The idea factory: Bell Labs and the great age of american innovation. Penguin Books, New York
[Glb:94] Gilb T, Graham D (1994) Software inspections. Addison Wesley, Reading
[Goe:31] Goedel K (1931) Undecidable propositions in arithmetic. Über formal unentscheidbare Sätze der Principia Mathematica und verwandter Systeme, I. Monatshefte für Mathematik und Physik 38:173–98
[Hea:56] Euclid (1956) The thirteen books of the elements, vol 1 (trans: Sir Thomas Heath). Dover Publications, New York. (First published in 1925)
[Hil:00] Hilzik MA (2000) Dealers of lightning. Xerox PARC and the dawn of the computer age. Harper Business, New York
[Hum:06] Hume D (2006) An enquiry concerning human understanding. Digireads.com, Stilwell. (Originally published in 1748)
[IGN:14] IGN presents: the history of Atari. March 2014. http://www.ign.com/articles/2014/03/20/ign-presents-the-history-of-atari
[Jac:99] Jacobson I, Booch G, Rumbaugh J (1999) The unified software development process. Addison Wesley, Reading
[Jac:05] Jacaobson I et al (2005) The unified modelling language, user guide, 2nd edn. Addison Wesley Professional, Upper Saddle River
[KaC:74] Kahn B, Cerf V (1974) Protocol for packet network interconnections. IEEE Trans Commun Technol 22:637–648
[Kan:03] Immanuel K (2003) Critique of pure reason. Dover Publications, New york. Originally published in 1781
[Ker:81] Kernighan B (1981) Why Pascal is not my favourite language. AT&T Bell Laboratories, Murray Hill
[KeR:78] Kernighan B, Ritchie D (1978) The C programming language, 1st edn. Prentice Hall, Englewood Cliffs
[Lam:72] Butler L (1972) Why Alto? Xerox inter-office memorandum. Xerox PARC, Palo Alto
[Lei:03] Wilhelm Gottfried L (1703) Explication de l'Arithmétique Binaire. Memoires de l'Academie Royale des Sciences 3:85–93
[Lov:42] Menabrea LF (1842) Sketch of the analytic engine invented by Charles Babbage. Bibliothèque Universelle de Genève, No. 82 Translated by Ada, Augusta, Countess of Lovelace
[MaP:02] Malmsten E, Portanger E (2002) Boo Hoo: $135 Million, 18 Months… A dot.com story from concept to catastrophe. Arrow, London
[Mc:59] McCarthy J (1959) Programs with common sense. In: Proceedings of the Teddington conference on the mechanization of thought processes. Her Majesty's Stationery Office, London
[McH:85] McHale D (1985) Boole. Cork University Press, Cork
[MeJ:01] Meurling J, Jeans R (2001) The ericsson chronicle: 125 years in telecommunications. Informationsforlaget, Stockholm
[Mor:65] Moore G (1965) Cramming more components onto integrated circuits. Elect Mag 38:14–117
[Mot:99] Motorola Museum of Electrics and Motorola (1999) Motorola (CB) – a journey through time and technology. Purdue University Press
[Nau:60] Peter N (1960) Report on the algorithmic language: ALGOL 60. Commun ACM 3(5):299–314

[Nes:56] Newell A, Simon H (1956) The logic theory machine. IRE Trans Inf Theory 2:61–79
[OGC:04] Office of Government Commerce (2004) Managing successful projects with PRINCE2. The Stationery Office, London
[ORg:06] O'Regan G (2006) Mathematical approaches to software quality. Springer, London
[ORg:10] O'Regan G (2010) Introduction to software process improvement. Springer, London
[ORg:11] O' Regan G (2011) A brief history of computing. Springer, London
[ORg:12] O'Regan G (2012) Mathematics in computing. Springer, London
[ORg:13] O'Regan G (2013) Giants of computing. Springer, London
[ORg:14] O'Regan G (2014) Introduction to software quality. Springer, Cham
[ORg:15] O'Regan G (2015) Pillars of computing. Springer, Cham
[Par:72] Parnas D (1972) On the criteria to be used in decomposing systems into modules. Communications of the ACM 15(12): 1053–1058
[Plo:81] Gordon P (1981) A structural approach to operational semantics, Technical Report DAIM FN-19. Computer Science Department. Aarhus University, Aarhus
[Por:98] Porter ME (1998) Competitive advantage. Creating and sustaining superior performance. Free Press, New York
[Pug:09] Pugh EW (2009) Building IBM: shaping an industry and its technology. MIT Press, Cambridge
[Res:84] Resnikoff HL, Wells RO (1984) Mathematics in civilization. Dover Publications, New York
[Rob:05] Robbins A (2005) Unix in a nutshell, 4th edn. O'Reilly Media, Sebastopol
[Roy:70] Royce W (1970) The software lifecycle model (Waterfall Model). In: Proceedings of the WESTCON, August, 1970
[SCA:06] Standard CMMI appraisal method for process improvement. CMU/SEI-2006-HB-002. V1.2. August 2006
[Sch:04] Schein E (2004) DEC is dead, long live DEC. The lasting legacy of digital equipment corporation. Barrett-Koehler Publishers, San Francisco
[Sch:14] Schaefer MW (2014) The Tao of Twitter. Changing your life and business 140 characters at a time, 2nd edn. McGraw-Hill, New York
[Sea:80] John S (1980) Minds, brains, and programs. Behav Brain Sci 3:417–457
[Sha:37] Shannon C (1937) A symbolic analysis of relay and switching circuits. Masters thesis, Massachusetts Institute of Technology
[Sha:48] Shannon C (1948) A mathematical theory of communication. Bell Syst Tech J 27:379–423
[Sha:49] Shannon CE (1949) Communication theory of secrecy systems. Bell Syst Tech J 28(4):656–715
[Sho:50] Shockley W (1950) Electrons and holes in semiconductors with applications to transistor electronics. Van Nostrand, New York
[Smi:23] Smith DE (1923) History of mathematics, vol 1. Dover Publications, New York
[Spi:92] Spivey JM (1992) The Z notation. A reference manual, Prentice Hall international series in computer science. Prentice Hall, New York
[Tur:50] Alan T (1950) Computing, machinery and intelligence. Mind 49:433–460
[Turn:85] Turner MD (1985) Proceedings IFIP cónference, Nancy France, Springer LNCS (201), September 1985
[VN:45] von Neumann J (1945) First draft of a report on the EDVAC. University of Pennsylvania, Philadelphia
[Wei:66] Joseph W (1966) Eliza. A computer program for the study of natural language communication between man and machine. Commun ACM 9(1)):36–45
[Wei:76] Joseph W (1976) Computer power and human reason: from judgments to calculation. W.H. Freeman & Co Ltd, San Francisco

Index

G. O'Regan, *Introduction to the History of Computing*, Undergraduate Topics in Computer Science, DOI 10.1007/978-3-319-33138-6

Printed in the United States
By Bookmasters